Iranian Air Force
From Establishment to the Beginning of the Jet Age

BABAK TAGHVAEE

AIR FORCES SERIES, VOLUME 14

Front cover image: Hurricane Mk.IIC fighters played an important role in Iran's war against Soviet-backed separatists in 1946. This image shows one of them before delivery during a test flight in the United Kingdom. (Author's archive)

Title page image: Mohammad Reza Shah Pahlavi modernised the Imperial Iranian Air Force (IIAF). He was a passionate aviator and became a pilot in 1946. He trained to fly with Hurricane Mk.IIC fighters and participated in combat missions during the operation to liberate Iran's Azerbaijan province from the hands of Soviet-backed separatists in 1946. (Author's archive)

Contents page image: This image, taken in the United Kingdom in 1934, shows '208', one of the Hawker Fury biplane fighter aircraft operated by the Imperial Iranian Air Force (IIAF) prior to delivery. It is powered by a Pratt & Whitney Hornet S2B1g radial piston engine, driving a three-bladed propeller. Due to the unreliability of these US-made engines, they were replaced with British-made 550hp (410kW) Bristol Mercury VISP radial piston engines, fitted with a two-bladed propeller. Hawker Furies remained the key fighter aircraft of the IIAF for 12 years until Hawker Hurricane Mk.IIC fighters took their place. Hawker Furies, together with Hawker Audax light-bombers, played an important role in the counter-insurgency (COIN) operations led by the Imperial Iranian Ground Force (IIGF) throughout Iran for almost 15 years. (Author's archive)

Back cover image: Iran's Junkers Company provided transport aircraft for the Iranian Armed Forces during the 1920s. This image shows a Junkers F 13 transport aircraft in use by the Persian Land Forces (Army Ground Force) during a counter-insurgency operation in Borujerd, west of Iran, in 1924. (Author's archive)

Published by Key Books
An imprint of Key Publishing Ltd
PO Box 100
Stamford
Lincs PE9 1XQ

www.keypublishing.com

The right of Babak Taghvaee to be identified as the author of this book has been asserted in accordance with the Copyright, Designs and Patents Act 1988 Sections 77 and 78.

Copyright © Babak Taghvaee, 2025

ISBN 978 1 80282 875 7

All rights reserved. Reproduction in whole or in part in any form whatsoever or by any means is strictly prohibited without the prior permission of the Publisher.

Typeset by SJmagic DESIGN SERVICES, India.

Contents

Introduction 4

Chapter 1   The Early Years 6

Chapter 2   The IIAF During the Years of Iran's Occupation 53

Chapter 3   Rebirth of the IIAF after World War Two..66

Introduction

Major General Reza Pahlavi, commander of the Iranian Army, took power in Iran and established the Pahlavi dynasty as a result of a coup d'état on 21 February 1921. Known as the founder of modern Iran, he became the first king, or Shah, of Iran. Over the next two decades, he oversaw reform throughout the country and modernised the Iranian Armed Forces. Under his rule, the Army was equipped with aircraft.

At that time, aviation was in its infancy, though it had been brought to the world's attention during World War One. Still, many countries, including some advanced nations, did not see the value of aviation for military purposes. Against this backdrop, the order to establish an aviation branch for the Iranian Army was issued on 1 June 1924 by Reza Shah during his tenure at the Ministry of War and as Commander-in-Chief. Since the Iranian army had to send its officers and staff abroad to be trained, the decree did not come into effect until they returned from France and Russia. Once the officers were trained, by March 1926, aviation in the Imperial Iranian Army could begin.

Despite its lack of equipment and infrastructure, including airports, as well as inadequate numbers of personnel, specifically pilots and technical experts, the aviation branch of the Army did help to resolve internal conflicts within the country, including suppressing rebels intent on overthrowing the central government. With security established in the country, the young Iranian Air Force, under the watchful eye of the government, progressed at extraordinary speed. With a broad plan to build aircraft in newly established factories, which were to be the future home of aircraft manufacturing in Iran, the Imperial Iranian Air Force was established. Operations commenced by receiving concessions from foreign manufacturing factories, after which a significant number of de Havilland DH-82 Tiger Moth training aircraft and Hawker Audax and Hind fighter aircraft were built in these factories.

Iran became involved in World War Two in August 1941 when Allied and Russian forces invaded the country, halting the nation's development plans for the Air Force and effectively threatening the existence of the IIAF. However, it survived, and progress was made with the support of Mohammad Reza Pahlavi, who replaced his father as Shah in 1941. However, by the end of World War Two, the Air Force had lost much of its combat and operational capabilities. Training of personnel stopped, the quality of aircraft maintenance was reduced, there were problems with technical support and the supply of spare parts for the aircraft, and these combined resulted in declining flight hours for the pilots, and reduced flight safety, leading to an increase in incidents and accidents.

Once Allied forces withdrew from Iran in 1946, the process of developing the Air Force began. Aircraft that had been ordered before the occupation of Iran were delivered. A large number of transport, training, fighter and liaison aircraft were also received as military aid from the United States, and these breathed new life into the Air Force. Although there were still shortages of aircraft until the late-1950s, the Air Force was progressing.

The withdrawal of American, British and Soviet forces from Iran in 1946 was followed by nationwide rebellions of Soviet-backed insurgents. The Imperial Iranian Army, supported by the Air Force's combat aircraft including Hawker Hurricane Mk.IIC fighters, played an important role countering them, preventing the Soviet Union from dissolving Iran and establishing a puppet regime.

Finally, in 1956, with the arrival of the first Lockheed T-33A Shooting Star training jets, the Air Force entered the modern jet age. A year later, the Air Force received its first fighter jets, Republic F-84G Thunderjet fighter-bombers. Despite these deliveries, the IIAF continued operating piston-engined fighters and transport aircraft for a number of years until the last Douglas C-47A/B Dakota transport aircraft was retired in 1971. In this book, the history of these old aircraft are reviewed.

In 1891, a French aviator flew a balloon over multiple cities in Iran, including Tehran. (Author's archive)

Chapter 1

The Early Years

The First Flights in Persia

In 1891, at the end of the reign of Naser al-Din Shah Qajar, a manmade flying device flew in Iran. A French aviator visited to demonstrate balloon flight in Tehran and other big cities including Tabriz. Yet, it was 23 years before another human-made flying device took to the skies over Iran.

Then, on 2 January 1914, just over ten years after the Wright brothers' flight at Kitty Hawk, the citizens of Tehran witnessed the first flight of an aircraft over the city. The aircraft was a Russian copy of the French-designed Blériot XI, known as the 'Rossiya-B'. It had been purchased by a Polish national from Russia. It arrived in Iran in kits through the port of Anzali in the Caspian Sea and after its assembly, it was flown from the Meydan Masqh (Training Ground) of the Persian Cossack Division, currently the location of the Iranian Ministry of Foreign Affairs. While landing, the aircraft collided with a barrel and was damaged although its pilot (known as Kuzminskii) was unharmed.

The aircraft was repaired in Tehran's military headquarters. Kuzminskii later used it to perform a number of display flights over Tehran, which were witnessed by Iran's king, Ahmad Shah Qajar. Despite these display flights, the Qajar government was not interested in buying aircraft for the Iranian Armed Forces from Blériot.

The first aircraft to land in Iran was a Russian copy of the French-designed Blériot XI known as the 'Rossiya-B'. In this image, taken on 2 January 1914, the aircraft, in the premises of the Meydan Masqh (Training Ground) of the Persian Cossack Division, had had its wings dismantled. (Author's archive)

Under the leadership of Vladimir Lenin and Leon Trotsky, the Bolshevik communists established the Soviet state on 7 November 1917, immediately after their October Revolution. Following the Russian Revolution in 1917, Marxists within Iran became more organised and began collaborating with the new ruling regime in Russia. The Soviets tried to spread Marxism into Iran using local rebels.

'Rossiya-B', taken in Tehran on 2 January 1914. (Author's archive)

'Rossiya-B', piloted by a Polish aviator Kuzminskii, over Tehran on 2 January 1914. (Author's archive)

In Iran's Gilan province, an Islamo-Communist insurgency named 'Jangal (Forest) movement' began under the leadership of Mirza Kuchik Khan. With Soviet financial and military support, he began an uprising with the intention of establishing the Socialist Republic of Gilan.

These rebels were confronted by the Persian Cossack Division under the command of the White Russians (a confederation of anti-communist forces) and were supported by the British Army. For the first time in Iran's history, during a joint counter-insurgency (COIN) operation of the British-Iranian-Russian forces against Mirza Kuchik Khan rebels, two Royal Air Force aircraft (of unknown type) were used for liaison and reconnaissance in Manjil on 12 June 1918. The operation ended successfully for Iran on 5 June 1921 and later, Kuchik Khan fled into the mountains, together with his companion named Gaouk, a Russian-German revolutionary adventurer. Both died of frostbite on 2 December 1921.

Ahmad Shah Qajar, the last king of the Qajar dynasty, is standing next to 'Rossiya-B' on 2 January 1914. (Library of Congress; George Grantham Bain Collection)

The Meydan Masqh (Training Ground) of the Persian Cossack Division in the centre of Tehran just a few years after the first aircraft landed there. (Author's archive)

The First Persians Fly Over Iran

Immediately after the coup d'état on 21 February 1921, Reza Shah considered the need to strengthen the Persian Cossack Division and transform it into a modern army so that it could suppress rebels in different parts of the country. The Ministry of War was ordered to purchase a number of aircraft to this end. Reza Shah's diary, published in October 1971, says:

> On 21 April 1921, I issued my order for the formation of a new Ministry of War. At that time, various tribes had started armed riots across the country and had disconnected various parts of the country from Tehran. In another area of Iran, Mirza Khuchik Khan had declared [the] Socialist Republic of Gilan. They [its people] were not obeying the central government and were refusing to pay taxes. To suppress their insurgency, we had to equip the army. We ordered some aircraft and we spent [our] last Shahi [national currency] for strengthening the army's power. It took three months until the Persian Army was ready to deal with any kind of insurgency across Iran.

In the years immediately after the coup of 1921, there are no official documents about events or developments of aviation in Iran, but according to contemporary newspapers, it is possible to understand how aviation evolved in Iran.

In its issue of 2 April 1922, the *Iran* newspaper reported about the flight of a Russian plane to Tehran via Anzali; then, in its issue of 16 April, it reports, 'The airplane that came to Tehran from Russia on 4 April was in the sky yesterday evening and undertook a display flight for the people of

Tehran.' Then, in the 24 April edition of this newspaper, a report indicated the first flight of Iranians over Tehran:

> Last Thursday (20 April 1922), the airplane that came from Moscow to Tehran took passengers for the first time for a tour in the air of Tehran. Prince Mohammad Hossein Mirza, Colonel Reza-Gholi Khan, Minister of Finance, and Monsieur Shumbansky, Russia's ambassador to Iran, flew over Tehran. After their flight, Brigadier Amanollah Mirza, Brigadier Morteza Khan, Brigadier Jaafargholi Khan and Colonel Mosaed Nezam from the Imperial Persian Army flew over Tehran. During the third flight, two Iranian parliament members, Tadayyon and Sheikhol Iraqin flew over the city together with army Colonel Mosaed Al-Doleh, for 20 minutes. The noise in the aircraft cabin was not high and they could speak together during the flight. The aircraft then left Tehran for Moscow yesterday.

Until the first half of 1922, there were no specific fields for aircraft to land and take-off in Tehran, but as *Iran* reports on 29 June 1922, 'Following Britain's request for the arrival of two aeroplanes in Tehran

RAF Hinaidi was an air base used by the RAF's Baghdad Air Command between 1922 and 1937. This base was home to two de Havilland DH.4 and two DH.9a aircraft and was also used to assist the armed forces under the command of Reza Khan Mir-Panj (later Reza Shah Pahlavi) to deal with the communist rebels led by Mirza Kuchik Khan in April 1924. The specific type of aircraft sent to Iran at that time is not known but it is likely that they were Airco/de Havilland DH-4 bomber/transporter aircraft. Image taken in 1924. (US National Archives and Records Administration)

and finding a place for them to land, the government has allocated the lands of Qaleh Morghi for this purpose and levelled it with the help of prisoners.' A few weeks later, on 1 August 1922, *Iran* published a story about the arrival of two British aircraft from Baghdad a day earlier and their landing at Ghaleh Morghi. The aircraft were later used to perform display flights over Tehran.

Formation of the Persian Military Aviation Organisation

Immediately after taking power, Reza Shah Pahlavi transformed the Persian Cossack Division, Persian Central Brigade, and Persian Gendarmerie into the Imperial Iranian Armed Forces, which was officially founded on 8 January 1922. It had 140,800 troops, of which 7,000 were from the Persian Cossack Division, 130,000 from the Gendarmerie, 11,800 from the Persian Central Brigade and 2,000 from the Ministry of War.

In 1922, the elements that made up the army were defined and took a fixed form at the end of the year. However, the Ministry of War faced several problems including the lack of trained staff officers. Two ministers of this new organisation, for the military units and the Supreme Military Council, were established by Military Decree No. 5, when Brigadier General Jahanbani and Colonel Shibani took the roles of head and deputy, respectively, of the military units. Jahanbani chose his members of the Army's General Staff.

In autumn 1922, the components of the general staff, including operations, adjustments, health, veterinary medicine, intelligence and stewardship, were created. Reza Shah centralised bases and troops in five major cities. According to General Decree No. 1 issued by Reza Shah, the Iranian Army was divided into five military regions, namely Tehran, Tabriz, Isfahan, Mashhad, and Hamedan. Each had an Army Division.

While the expansion of the Army Land Forces was fast, the Air and Naval forces grew at a slower pace since, as well as equipment, they also required highly trained personnel.

On 4 June 1923, the first group of Army officers was selected to be sent to France for military training and education. One of the officers was Colonel Ahmad Khan Nakhjavan from the Infantry Guard Brigade, who chose to study aviation and become a pilot. He later became the first Iranian pilot to graduate. On 6 January 1924, he logged his first solo flight. In June of that year, ten more pilots were sent to Russia for training. Three of them, Sharafoldin Mirza Qahramani, Isa Khan Ashtodakh and Ali-Akbar Amin-Zadeh, completed their training in early 1926 and together with Nakhjavan became the first four pilots of the Iranian Army Aviation Organisation (later the Air Force).

Colonel Ahmad Khan Nakhjavan, the first Iranian pilot, is stood next to his aircraft, an Avro 504K, with his French instructor in Istres, France, following his solo flight on 6 January 1924. He was later appointed as one of the first commanders of Iran's Military Aviation organisation. (Author's archive)

Left: Isa Khan Ashtodakh was one of the first three Iranian pilots trained in Russia who became R-1 bomber pilots in 1924. He later became a test pilot with the IIAF and flew 69 different types of aircraft over 3,000 flying hours. After his service in the IIAF, he was appointed as the head of the Iranian Civil Aviation Organisation between 1952 and 1958. He passed away in Tehran in 1993. (Author's archive)

Below: Fifteen years after the first Iranian men became pilots, Iranian women also started training to become pilots. By order of Reza Shah, the Imperial Iranian Aeroclub was established in 1939 to train civilian pilots. In line with his gender equality policies, Iranian women were encouraged to become pilots, and subsequently, Effat Tejaratchi, Ina Avshid and Qudsieh Farrokhzad became the first Iranian women to enrol in the club on 27 September that year, and a few days later made their first solo flights. This photo shows Ina Avshid, after her first solo flight in the de Havilland DH.82 TigerMoth in Tehran. (Author's archive)

First Aircraft of the Persian Military Aviation Organisation

A Soviet-registered Junkers F 13, R-RECA from Junkers-Luftverkehr, had visited Tehran in April 1923. It was evaluated by the Iranian Armed Forces and by the Persian (later Iranian) government. Subsequently, the government showed interest in acquiring six. Their acquisition took place at the same time as the Aviation Office of the Persian Armed Forces was established. The Junkers F 13s were operated by the Junkers-Luftverkehr for the Iranian government to carry passengers and parcels. One of the aircraft was dedicated to the military.

In addition, a second aircraft was procured exclusively for military use. It was an R-2, a Soviet copy of the Junkers A20, a two-seater postal, training and military aircraft. The aircraft, equipped with an L-5 piston engine producing 300hp, could fly at a maximum speed of 125mph (201km/h) while its cruise speed was 100mph (161km/h). It was named 'Mazandaran', while the Junkers F 13 in use by the Iranian Army was named 'Gilan'. These two aircraft were purchased with money collected by the 'Mazandaran' and 'Gilan' Brigades in the north of Iran.

At 15:00 local time on 26 May 1924, the newly procured Junkers aircraft were displayed at Qaleh Morghi airport, Tehran. Brigadier General Reza Pahlavi (later Reza Shah Pahlavi), the head of the Iranian National Parliament, parliament members, army commanders and local people were present. In total, six aircraft were flown that day and all were piloted by non-Iranian pilots. The sole Junkers F 13 and the sole A20 was among them and carried army generals as passengers. After the first round of display flights, the aircraft landed and picked up a number of others including journalists. On the same day, other Army aircraft were deployed to Broujerd in the west of Iran.

In July and August 1924, two RAF DH.9As visited Tehran. The local authorities were invited to see demonstrations but none of the officials accepted as the Iranian government was pursuing an anti-British policy in fear that its country would be colonised again.

To equip the Army with combat aircraft, Reza Shah's government requested the acquisition of military aircraft from the United States government, a country it considered neutral, on 17 January 1923. This request was rejected. At the time, the US government was pursuing a disarmament policy following decisions made during the Washington Naval Conference held in Washington, D.C. from 12 November 1921 to 6 February 1922.

Finally, on the order of General Amanullah Mirza, Chief of Staff, and with the approval of Reza Shah, the first six Persian Air Force aircraft were ordered from France, a country also considered 'neutral'. The aircraft were two Breguet 14 two-seat reconnaissance-bombers, two Potez VIII training aircraft and two Spad 42 training aircraft. In 1923, several Airco DH.4 and DH.9a (Polikarpov R-1) were also ordered from the Soviet Union. The Polikarpov R-1 was a Russian-built under license version of the British Airco DH.9a powered by a Russian-made 400hp Liberty piston engine.

The French aircraft arrived in Iran through the port of Bushehr on 25 January 1924. After their assembly, they were flown to Tehran by pilots Berhault (chief of mission) and Gérard, and two mechanics, Banel and Sigann. One of the six aircraft crashed en route leaving five for the newly formed Air Force. The first of the five to reach Ghaleh-Morghi airport, Tehran, was a Breguet 14 on 22 May, while the other Breguet 14, which had crashed, arrived by lorry. After repair and refurbishment, the five aircraft entered the Air Force service as 1 *Taghdir*, 2 *Oghab*, 3 *Shahin*, 4 *Homa* and 5 *Simorgh*.

Two DH.4s under the control of two Russian pilots, Vasil'chenko and Lozovsky, had also arrived in Tehran on 12 May 1924. Later, three Soviet-made DH-9as (Polikarpov R-1s) were flown to Iran with a refuelling stop in Baku on 26 January 1926. These three were flown by the first three Iranian pilots who were trained in Russia. They were First Lieutenant Qahramani, First Lieutenant Amin Zadeh and Student Pilot Isa Ashtodakh. Amin-Zadeh's R-1 was damaged during take-off from Baku and was dismantled and sent to Iran in parts. Qahramani's R-1 was damaged during landing in Qazvin, near the city of Tehran, and was later sent to Qaleh-Morghi by road to be repaired, while Ashtodakh's R-1 was grounded in Qazvin for several days due to heavy snowfall and bad weather. It later joined four others in Ghaleh-Morghi in February 1926.

The first French-built aircraft that was flown and ferried to Iran by an Iranian pilot was a Bréguet 19 and reached Qaleh-Morghi airport on 24 February 1926. The aircraft, piloted by Colonel Nakhjavan, began in Villacoublay airport, Paris, with refuelling stops in six cities including Istres, Venice, Eskişehir, Aleppo and Baghdad. Reza Shah Pahlavi arrived at Ghaleh-Morghi airport to witness the aircraft landing. Nakhjavan became the first Iranian pilot who landed in Ghaleh-Morghi airport. The other Iranian pilots had landed their Russian R-1 aircraft in Qazvin and were unable to transfer them to Tehran due to a snowstorm and other bad weather.

Colonel Ahmad Nakhjavan had earlier been appointed as the head of the Iranian Military Aviation Organisation. On 22 March 1926, Reza Shah ordered the expansion of the organisation to become the Aviation Department and kept Nakhjavan as its commander. At that time, the department had only five Vickers machine guns, four Lewis automatic machine guns, three bomb racks for Bréguet-19 aircraft, four Lewis machine gun magazines, one bomb rack and three bomb ejectors for R-1 aircraft.

In March 1926, the Aviation Department of the Iranian military had just 20 aircraft and among them 12 were airworthy. These aircraft were a Bréguet 19 with serial number (s/n) 1; two Bréguet 14s with s/n 2 (under maintenance) and 3 (active); a Russian DH.4 with s/n 4 (active); five Russian DH.9as with s/n 5 to 9, among these number 7 was withdrawn from use due to an incident, number 8 (named *Oghab* or 'Eagle') was under maintenance, while the remaining three were airworthy; two Junkers A20s with s/n 21 and 22; three Junkers F 13s with s/n 23 to 25; two Spad 42s with s/n M-1 and M-2 (both grounded awaiting maintenance); two U-1s (Soviet copies of Avro 504Ks) with s/n M-3 (active) and M-4 (withdrawn from use); and two Potez VIIIs with s/n M-5 and M-6 (both grounded awaiting maintenance).

On 1 May 1926, the first airshow of the Iranian Military Aviation Department was held at Ghaleh-Morghi airport and involved nine aircraft, which were piloted by five Iranian officers, two Russian instructor pilots, a Russian pilot and a German instructor pilot. Reza Shah, a group of government and military officials and tens of thousands of people from Tehran were present to witness the airshow. Some Air Force officers were trained to perform wing walking on the R-1s at the airshow.

Colonel Ahmad Khan Nakhjavan in France, this time next to a Bréguet 19 sesquiplane bomber and reconnaissance aircraft on 24 February 1926, just before he flew with the aircraft to Iran. It was the first Iranian military aircraft to land in the country under the control of an Iranian pilot. It later received s/n 1 and its pilot, Ahmad Nakhjavan, was appointed as head of the Persian Military Aviation Organisation. (Author's archive)

As the budget of the newly formed air power was extremely limited, a special decree was issued on 21 April 1926 to reduce pilot flight time to a maximum 40 hours per year to save on the budget for oil and fuel. Each pilot was allocated a maximum of three hours and 20 minutes flying time per month. Despite this, each Iranian pilot, until 21 April 1927, had 46 hours and 24 minutes flight time each year.

At Ghalel Morghi airport, Tehran, is a Junkers F 13 transport aircraft during Iran's first airshow. Women can be seen fully covered in Islamic veils. On 8 January 1936, Reza Shah issued a decree known as Kashf-e hijab, banning all Islamic veils (including hijab and chador), an edict that was swiftly and forcefully implemented. It was part of his efforts to bring gender equality to Iran and gave freedom to Iranian women. (Author's archive)

Taken in Tehran's Ghalel-Morghi airport in 1926, this photograph shows one of the first R-2 (Soviet copy of Junkers A20) training and observation aircraft in use by Iran's Military Aviation during Iran's first airshow. (Author's archive)

Walter Mittelholzer, a Swiss aviation pioneer and photographer, flew all over Iran with Junkers F 13 from Iran's Junkers Company to undertake aerial photography. One of these F 13s (R-RECI) was used for passenger and parcel transport in Iran, is over Tehran in 1926. (Walter Mittelholzer, ETH-Bibliothek's collection)

One of the most successful early airliners at the beginning of the 1920s was the all-metal Junkers F 13 monoplane. The aircraft became more popular in the civilian market than with the Iranian Armed Forces. Iran's Junkers Company used them to carry passengers and parcels all over Iran, including to Bushehr, where one of them can be seen in 1927. Junkers initiated a subsidiary firm in Persia known as Junkers Luftverkehr Persien in 1927. The line started on 8 February of that year with a flight from Tehran to Bandar Pahlavi. Later, a line was added from Tehran to Bushehr / Baku (in the Soviet Union). The Junkers firm closed in 1932 due to the world economic crisis. (Kees Kort Collection)

A rare image taken in Tehran in 1926 shows two Junkers F 13 passenger/transport aircraft and two R-1 bombers from the Persian Military Aviation during Iran's first airshow and prior to the visit of Reza Shah on 1 May 1926. (Kees Kort Collection)

Two Junkers F 13s operated by the Persian Military Aviation during Iran's first airshow at Ghaleh Morghi airport on 1 May 1926. The aircraft lacked any registration codes or serial numbers when this image was taken. (Author's archive)

First Operations of the Persian Military Aviation Department

The first use of aircraft for military purposes in Iran took place on 12 June 1918 when a pair of Royal Air Force aircraft of unknown type were used for liaison and reconnaissance in support of an Iranian counter-insurgency operation (COIN) against an Islamo-Communist group in Manjil, northern Iran. It was several more years before the Iranian armed forces used aircraft in their operations. On 14 November 1924, the newly formed Iranian Military Aviation Organisation (which later became the Air Force) used its Junkers A 20 and F 13 aircraft under the control of German pilots in support of a COIN operation against armed separatists led by Sheikh Khazaal in the Khuzestan province of Iran.

To encourage the rioters to surrender, the 1st Infantry Division of the Iranian Army dropped leaflets all over the cities of Khuzestan province. In the text on the leaflets, the Iranian Army gave assurances to the local population that its operation was against Sheikh Khazal (Khaz'al Ibn Jabir) rebellion.

A few days later, other military aircraft were used in support of another counter-insurgency operation, carrying out reconnaissance and liaison activities during the rebellion of Bakhtiari tribes in Isfahan province of Iran.

The COIN operation in Khuzestan province ended with defeat of the separatists and the liberation of Khorramshahr on 2 December 1924, while the operation against Bakhtiari rebels ended with similar victory a few days later. During these operations no bombs were dropped by the Iranian military aircraft. Their weapons were, in fact, leaflets, which contained psychological warfare content, breaking the morale of the separatists.

Until June 1926, none of the Iranian pilots had combat experience. In that year, the whole country, especially in Azerbaijan, Balochistan, Kurdistan and Lorestan, faced insurgency and crimes. To establish the rule of law and security in these provinces, the Army launched several new COIN operations. Two of these operations involved the Aviation Department and its aircraft; one in Salmas in the north-west and the other in Maraveh-Tappeh in the north-east of Iran against insurgents, both starting on 24 June 1926. Another operation against Kurdish rebels in Sanandaj was launched a few days later.

On 2 July 1926, three aircraft, comprising a Bréguet-19 piloted by Colonel Ahmad Nakhjavan, an R-1 piloted by 2nd Lieutenant Isa Ashtodakh and a Junkers F 13 piloted by a German contractor carrying three mechanics (French, Russian and German), flew to Mashhad to fly attack missions against insurgents in Maraveh-Tappeh.

On 22 August 1926, two other aircraft, a U-1 under the control of an Iranian pilot and an R-1 under the control of a Russian pilot, were deployed to Sanandaj to provide close air support for the Army forces involved in another COIN operation. The R-1 crashed due to engine failure during take-off at Sanandaj on 31 August 1926, while the U-1 crashed at Ghaleh-Morghi airport due to engine failure after returning to Tehran on 4 September 1926.

The third involvement of the Aviation Department in the COIN operations began on 7 September 1926 after three aircraft under the control of two Iranian pilots and one German pilot were deployed to Tabriz. They were two Bréguet-19s and a Junkers F 13 and took part in the COIN operation in Maragheh. One of the Bréguet-19s was grounded due to engine failure and a replacement aircraft, a Bréguet-14, was selected to be deployed to Tabriz in its place, but while it was landing at Qazvin for refuelling on its way to its destination, its landing gears hit a telegraph wire and were damaged. It was grounded in Qazvin then ferried back to Tehran by a French mechanic to be repaired. A similar landing gear mishap happened to the Junkers F 13 in Tabriz, resulting in it being grounded for several days.

Until 1928, when eight Iranian pilots graduated from two flight schools, one in Istres, France (four pilots), and another in Moscow, USSR (four pilots), the Aviation Department of the Imperial Iranian Army suffered due to a lack of pilots. Subsequently, foreign pilots were contracted to fly aircraft alongside Iranians. For example, in October 1926, when all 12 airworthy aircraft were on deployment, the Iranian Army asked for one more aircraft to be deployed to Tabriz to be used in support of its COIN operation

This image taken in 1926 shows one of Iran's Junkers A20s before its involvement in a military operation in Tabriz. (Author's archive)

Iranian airmen stand next to their Junkers A20 and R-1 aircraft prior to their participation in a military operation in Sanandaj in 1926. (Author's archive)

in Salmas. A recently repaired Junkers A20 was the only flyable aircraft available in Qaleh-Morghi to be deployed. The Army contracted a civilian German pilot named Scheffer, from Iran's Junkers company, to fly it. After the operation, Colonel Nakhjavan, 2nd Lieutenant Ashtodakh and Scheffer were awarded the 3rd Degree Order of Sepah Medal for their bravery during the operations.

As well as a shortage of pilots, the Aviation Department also suffered due to a lack of Iranian technicians, requiring the department to employ 30 French, German and Russian mechanics to maintain and repair its aircraft. As no observers or aviation photographers served with the Iranian Army, foreign mechanics were instructed to do the job and in rare cases fire machine guns and air-drop bombs.

Continued Insurgencies and Difficulties for Ground Units

In September and October 1927, three more Russian-built DH.9a (R-1) aircraft were procured from the Soviet Union for the Aviation Department. They were delivered by ship through the Caspian Sea port of Pahlavi (Anzali). They were assembled there and then were flown to Tehran by Iranian pilots. Each had a Vickers machine gun, a machine gun turret and three bomb racks. Despite the increase in the number of aircraft and Iranian pilots, the Aviation Department was still under pressure as the number of COIN operations increased in various parts of the country, particularly in mountainous regions.

The COIN operations against Bakhtiari insurgents in the Lorestan region of Iran had brought difficulties for the government. The distance between Borujerd airport and the area of the operation had caused difficulties for a pair of the aircraft that had been deployed there to carry out close air support (CAS) missions in 1926. In 1927, decisions were made to use a smaller airport at Khorramabad instead, which significantly increased the time to target of the aircraft during CAS missions. These operations had two Junkers A20s and four R-1s (DH.9a copies) involved. They were piloted by four Iranian officers and three non-Iranian contractor pilots (one from Germany and two from the USSR).

The Junkers A20 two-seater cantilever monoplane was always preferred for operations in the Lorestan district of Iran. One of them, under the control of a German pilot and with a Russian mechanic seated in the back to control a flexible 7.7mm (0.303in) Lewis gun) experienced technical failure and crashed on 29 January 1928, resulting in injuries to the crew. The German pilot was the first war-wounded aviator in Iran's military aviation.

The Iranian Armed Forces lost eight of its aircraft over two years. They were a DH.4, four DH-9As (all Soviet R-1s), a Junkers F 13 and two U-1s. In October 1928, the Aviation Department under command of Brigadier Ahmad Nakhjavan (who was promoted on 21 March 1928) had 24 aircraft, which were in use with three squadrons. The 1st Squadron had seven R-1s single-engined light biplane bombers; the 2nd Squadron had two Soviet DH.9as (R-1s), two Junkers F 13s and two Junkers A20s; while the Combined Training Squadron (third unit) had 11 aircraft, comprising a Bréguet-14, a Bréguet-19, five U-1s (Soviet copies of Avro 504Ks), two Potez VIIIs and two Spad 42s.

In summer 1928, while the Iranian Armed Forces had managed to establish security and control over north-west, north-east and west Iran, a new challenge began in south-east Iran. In the Baluchistan region, leaders of Baluch bandit gangs started an insurgency against the Pahlavi government, which forced the Imperial Iranian Army to launch a COIN operation against them, and called on military aircraft to be involved from summer 1928. The aircraft were forward deployed to Zahedan, with refuelling stops at Gorgan and Mashhad.

From 28 October to 28 December 1928, three R-1s were also deployed to south-west Iran to take part in another COIN operation. They were flown by Iranian pilots, while their observers and gunners were foreign mechanics hired by the Iranian Army.

Nomads of Isfahan and Fars regions of Iran were provoked by local landlords to start an insurgency against the central government to avoid paying taxes. This action prompted the Pahlavi government to

For years, Junkers F-13s acted as the key means of transport for the Persian Military Aviation Organisation. One of them can be seen during a military operation in Borujerd in 1924. (Author's archive)

launch two COIN operations against them, forcing them to surrender in 1929. To allow the Aviation Department to fly its aircraft on CAS missions in support of the Army, Shiraz's airport was chosen as its forward operating base during the operation between 18 February and 14 March 1929.

As the COIN operations were increased and more military aircraft became involved, more pilots and technicians were needed to fly them. In 1928, two groups of the Army cadets were sent to the Soviet Union, the first consisting of three cadets who attended Sevastopol's flight school for pilot training starting from 12 May 1928 and the next group of eight being sent to Leningrad's technical school to pass aircraft maintenance training from 22 September 1928. This increased the number of pilots in the Iranian Armed Forces to 15. Among these, six Iranians had completed training in Istres, France, while eight other Iranians and one Soviet national were graduates of Kacha Military Aviation School (later Kacha Higher Military Aviation Twice Red Banner Order of Lenin School of Pilots named for A.F. Myasnikov) in Sevastopol, Soviet Union.

Increasing Operations in the South

With the rise of insurgencies in the south of Iran, the Iranian military, which had now been relieved of operations in the north-west and north-east, focused on dealing with nomads who had been used by wealthy landowners and become known as bandits against the central government. Shiraz's airport played an important role in these operations as a forward operating base, the Aviation Department flying its aircraft from there to other regions to carry out missions. For example, on 15 May 1929, the Aviation Department had two Russian DH.9as that were on deployment in Shiraz flown to the south of Iran.

The leader of the formation was 3rd Lieutenant Fathollah Zand-Ansari, while the second was a Soviet contractor pilot named Lyakhovsky. Their technicians were both Russian.

During one of the missions performed by Zand-Ansari, the engine of his R-1 aircraft failed and he had to perform an emergency landing in an area that was under the control of insurgents in Ardakan, near the city of Yazd. The pilot and his observer (back seater), 1st Lieutenant Mahmoud Mirza-Khosravani, were not injured. The insurgents saw and heard the aircraft prior to its emergency landing, so, to avoid having their aircraft captured, the pilot and his team set it on fire. The Lewis and Vickers gun were destroyed, so the insurgents believed the pilot died in the aircraft. Their action bought them enough time to reach Shiraz, a walk of almost 300km (186 miles).

Just a few days after returning to Shiraz, Zand-Ansari and Mirza-Khosravani came under attack as insurgents raided the airport where the Aviation Department had three other R-1s on deployment. They opened fire at the aircraft and the building where Zand-Ansari and other officers were located. Three bullets hit Zand-Ansari's helmet; he was slightly injured but survived. All three R-1s were damaged and grounded for days until repaired. As the city of Shiraz was almost besieged, replacement aircraft were deployed and saw extensive use on CAS missions against the armed nomads to break the siege.

On 17 June 1929, eight more Army cadets were chosen to become pilots. Under supervision of an officer named Sultan Mahmoud-Khan Baharmast, they were sent to Istres, France, for pilot training.

On 6 July 1929, the Aviation Department received two new aircraft, a pair of German-made Junkers W 33 single-engine, low-wing monoplane transport aircraft. They were ferried to Iran by two German pilots who were contracted to fly them for the army. They received s/n J-7 and J-8 and were quickly deployed to Shiraz to be used in the COIN operations in the south. Together with eight Iranian pilots, eight non-Iranian mechanics and several Iranian observers, they participated in armed patrols and CAS missions in support of the Army's COIN operation in Fars and Isfahan regions until 1931, when the operation ended with victory.

More Polikarpov R-1 for the Iranian Armed Forces

Among all aircraft used by the Imperial Iranian Armed Forces during the COIN operations, Polikarpov R-1s, the unlicensed copies of British Airco DH.9a bombers had the best results. These Soviet light bomber and reconnaissance aircraft were powered by the M-5 copy of the American Liberty engine. More than 2,400 were built from 1922 to 1932. They were more powerful than the original British variant and were relatively cheaper to buy. The DH.9a and R-1 were visually identical to each other; however, the latter was built with material the Soviets had to hand locally. It had a wooden structure, built of pine. The fuselage was covered with plywood, apart from a small fabric section towards the rear. The wing spars were made of pine and the interplane struts had ash leading edges.

GAZ No.1 factory (previously known as Dux factory) in Moscow manufactured DH.9a copies for Iran. According to its documents, up to 1927, the Imperial Iranian Armed Forces received 26 R-1s. In January 1927, ten more were sent to Iran, among them five had original Liberty engines (manufactured in the United States) while the rest had M-5 copies. Two had gun turrets (TOZ turret) for the back-seater while the rest are claimed by the Russians to have Maxim machine guns in the back. In 1929, deliveries of the R-1 were continued. Thus, on 15 August, the representative of the Persian Ministry of War, Isa-Khan Ashtodakh, accepted from GAZ No. 1 four R-1s with construction numbers (c/n) 3862, 3863, 3864 and 3867, and in total six aircraft were delivered in 1929. The following year, six more R-1s arrived in Persia.

According to documents of the Imperial Iranian Air Force (IIAF), ten R-1s were ordered in 1929. On 24 February of that year an Iranian officer was sent to Moscow to prepare the paperwork for procurement and delivery of the aircraft. On 20 August the first of these aircraft was ferried by the same Iranian officer from Moscow to Tehran. The flight duration was 25 hours and 55 minutes and included six refuelling stops. The aircraft reached Tehran's Ghaleh-Morghi airport on 23 August 1924. Three other aircraft were delivered in parts (disassembled) across the Caspian Sea. They were assembled at Pahlavi port and were flown to Tehran by Iranian pilots on 22 September.

On 1 April 1930, another Iranian officer was sent to Moscow to receive the other six aircraft, which according to the IIAF's documents, were sent to Iran by ship through the Caspian Sea in June 1930. After assembly and test flights at Pahlavi port, they were flown to Tehran by Iranian pilots. Their delivery increased the total number of aircraft in use by the Aviation Department (Edareh Havapeymaei) to 33. They were Bréguet-19 (one), 17 Polikarpov R-1s (one under maintenance and two for pilot training), one Junkers F 13, two Junkers W33s, two Junkers A20 (both for pilot training), one Bréguet-14 (for pilot training only), five U-1s (all used for pilot training and among them two were under maintenance), two Spad 42s (both under maintenance) and two Potez VIIIs (both had brand new engines but fatigued fuselages).

The R-1 was the key aircraft in use with the Iranian Armed Forces and the type was heavily engaged in the second round of COIN operations in the south of Iran from early July 1930. At the start of the operation, once again the Iranian Armed Forces suffered from lack of pilots for the R-1s until nine Iranian cadets (three trained in the USSR and six in France) completed their pilot training. They began their flights on the R-1 during the operation from July–August 1930.

The second COIN operation in the south employed seven R-1s piloted by Iranian officers and one W33 piloted by a German contractor from Shiraz airport. The greatest number of operations took place in August and September 1930. By mid-February 1931, the operation ended with victory. Four of the R-1 pilots were promoted as a result of their participation in the operation.

Brigadier Ahmad Nakhjavan, commander of the Aviation Department was promoted to Brigadier General by Reza Shah Pahlavi and appointed as the head of the Imperial Iranian Army's General Staff on 18 August 1930. As a temporary replacement for him, Major Ahmad Mirza-Khosravani was temporarily appointed as commander of the Aviation Department. Under command of Mirza-Khosravani, the aviation element of the Iranian Armed Forces was restructured with three squadrons. These were 1st and 2nd Observation Squadrons as well as the Training Squadron. The 1st Observation Squadron consisted of two groups, the first with two units and second with one unit; while the 2nd Observation Squadron had two groups each with one unit. On 31 December 1930, Brigadier Ahmad Nakhjavan was returned to his position as the head of the Aviation Department.

A U-1, which is a Russian-built version of the Avro 504K Dux training aircraft. In addition to the pilot training, the U-1s were used for reconnaissance missions in support of larger aircraft such as R-1s and DH.9as during military operations. (Author's archive)

R-1 was the Russian copy of the Airco DH.4 and its more powerful variant, DH.9a. In the early years of Iran's military aviation, R-1s were widely used on COIN operations. An R-1 is in service with the Red Army Air Force. (Author's archive)

The majority of Polikarpov R-1s in service with the Iranian Armed Forces were DH.9a variants. The De Havilland/Airco DH.9a was a development of Airco's earlier successful DH.4, with which it shared many components. This image shows a Westland-built Royal Air Force DH.9a. (Author's archive)

In 1929, the Persian Military Aviation received two Junkers W33 transport aircraft, which were the largest aircraft in Iran until the last example was withdrawn from service due to lack of spare parts in 1939. This image shows one of them during an official visit of Reza Shah from Ghaleh-Morghi airport in 1929. (Author's archive)

Reza Shah is standing next to an R-1 operated by the Persian Military Aviation Organisation in Tehran's Ghaleh-Morghi airport. (Author's archive)

An Iranian R-1 possibly taken in Tehran in the early 1930s. (Author's archive)

A group of Iranian and Russian R-1 aircraft crew are standing next to an aircraft, which is parked in front of the first aircraft hangar at Ghaleh-Morghi airport, Tehran. The R-1, painted in a dark brown colour scheme, has an Iranian flag and a large lion and sun emblem on its vertical stabiliser. (Babak Taghvaee)

Accidents and Incidents

Between 28 April and 14 August 1930, four Imperial Iranian Armed Forces aircraft crashed. They were a U-1, a Junkers F 13 and two R-1s. Between 1 September and 10 November 1930, three more aircraft comprising a U-1 and two R-1s crashed. None of the crashes were fatal but they raised concern about flight safety and resulted in the formation of a crash investigation commission or team within the Aviation Department. Established on 23 November 1930, it had a remit to investigate the causes of crashes in order to prevent the same from re-occurring.

The first crash investigation team had five members, comprising commanders of the 1st and 2nd Observation Squadrons, a representative of the maintenance group, the commander of the group that suffered the crash and the pilot involved in the crash.

One of the Iranian aircraft mechanics of the Aviation Department, 2nd Lieutenant Siavash Siahpoush, built an R-1 using components of several crashed examples while at Qaleh Moghri airport in 1930. He built the wing and fuselage and used components including an M-5 piston engine. For his efforts for building an R-1 bomber, he was awarded 1,000 Iranian Rials and was promoted. All test flights were performed by 1st Lieutenant Mahdi Sepah-Pour while Siahpoush was seated in the back of the aircraft. During the second test flight, Reza Shah witnessed the aircraft flying. After the test flights, the R-1 received s/n 15-15 and joined the 1st Observation Squadron.

2nd Lieutenant Siavash Siahpoush rebuilt an R-1 using parts from crashed R-1s. (Author's archive)

Domestic Pilot Training in the de Havilland DH.82s

With lessons learned from the operations of 1928 to 1930, the Imperial Iranian Armed Forces began domestic pilot and aircraft mechanic training in Iran. A Swedish captain named Elis Nordquist was hired as advisor to the Aviation Department to establish a pilot and a mechanic training school in Ghaleh-Morghi in January 1932. Several training aircraft were evaluated by him and the Iranian Armed Forces and included de Havilland DH.82 Tiger Moth training aircraft, which Nordquist, together with Lieutenant Isa-Khan Ashtodakh and Major Arfa, evaluated in June 1932, and finally ordered for use as future training aircraft.

The Tiger Moth biplane was one of the best pilot trainers of its time, and soon became the key pilot trainer of the Iranian Armed Forces. In total, 20, with construction numbers (c/n) 3117 to 3136 were built on the order of the Iranian Ministry of War. They were shipped to Khorramshahr port from the UK in November 1932. Following assembly and a test flight by de Havilland's test pilots, they were flown to Qaleh Morghi airport by Iranian pilots.

Ferry flights to Tehran began on 3 December 1932. The aircraft flew in four five-ship formations to Tehran until 23 January 1932. On the way to Tehran, one aircraft crashed and was written off. Its pilot, 1st Lieutenant Fathollah Zand-Ansari, lost his life. He became the first Iranian pilot to lose his life in an aircraft crash.

A parade was held on 22 February 1933 involving 18 of the recently delivered DH.82s, which were flown from Ghaleh-Morghi airport. Two aircraft had a mid-air collision in front of Reza Shah Pahlavi, resulting in the death of 1st Lieutenant Qaem-Maghami, while the other pilot, 1st Lieutenant Fathi, was severely wounded. This accident angered and upset Reza Shah, who subsequently ordered the Iranian Armed Forces to improve flight safety regulations to avoid similar tragic events.

In September 1931, an engineering school was established in Tehran to train future aircraft mechanics at Ghaleh-Morghi airport. In total, 30 volunteers passed the entry exams and started their training on 17 October 1931. Among these, 23 successfully graduated in July 1933.

The first 20 applicants who successfully passed their exams began pilot training in September 1932. Of these, 14 students completed their training and were promoted to the rank of 2nd Lieutenant in July 1934. In September 1935, the second pilot-training course that had begun in 1933 also concluded, with 18 cadets graduating. From 1935, the number of cadets attending pilot-training courses gradually increased.

Despite the accidents and the aircraft losses, DH.82s were successfully used to train 14 Iranian pilots in Tehran in 1933. Several Swedish instructors led by the Swedish Air Force's instructor pilot, Captain Erik Ekman, were contracted to train Iranian cadets until existing Iranian pilots could be trained as instructors to take on the role.

To increase the number of training flights, nine DH.82As were ordered (c/n 3200 to 3208) in early 1933, followed by an order for another 14 (later reduced to ten aircraft, with c/n 3228 to 3237). Unlike the first 20 aircraft, which had Gipsy III piston engines with a maximum 120hp, these had Gipsy Major piston engines with 130hp maximum power output.

The nine DH.82As ordered in early 1933 were delivered in December and were flown to Tehran after assembly at Ahvaz airport while the ten aircraft ordered in September were delivered in September 1934. This increased the total number of DH.82s to 27.

To train more pilots, four Swedish instructors were recruited. They were 1st Lieutenant Frederik Johannes, 2nd Lieutenant Stik Søndgren, 2nd Lieutenant Kenar Krinos and 3rd Lieutenant Henrik Skolin; they arrived at Tehran on 21 January 1934. On 30 May 1934, a DH.82 under the control of Henrik Skolin crashed. Due to substantial damage (75 per cent), it was withdrawn from service. The crash investigation showed that human error was the cause of the crash, which resulted in Skolin's dismissal. On 4 July 1934, Reza Shah dismissed the rest of the Swedish instructor pilots. Nordquist also left Iran for Sweden. A new advisor (Colonel Victor Bevir from the Belgian Air Force) was hired as a replacement and as an air advisor. Alongside, six Belgian Air Force instructors were hired and trained Iranian pilots until August 1936.

The DH.82 was not the only aircraft type used for pilot training at Ghaleh-Morghi. In late 1932, the Iranian Ministry of War ordered ten Polikarpov R-5 reconnaissance bombers, which were delivered from 28 May 1933. The first was flown by 1st Lieutenant Ebrahim Makoei from Moscow to Tehran. The remaining nine were delivered by ship to Pahlavi port, where they were assembled and flown to Tehran by Iranian pilots. On 30 May 1933, the first R-5 reached Tehran after refuelling stops in Kharkov, Grozny and Baku.

The R-5s entered service with the 1st Regiment, being used for advanced pilot training. However, their Mikulin M-17B V-12 liquid-cooled piston engines of 507kW (680hp) were found to be unsuitable for operations in Tehran during the hot summer, resulting in multiple incidents.

In May 1933, two pilots, five aircraft maintenance officers and three recently graduated mechanics were sent to the UK for complementary education and training to become future instructor pilots and mechanics.

In December 1933, a third school was established for air observer training. Its first group of 18 students graduated in October 1934 and were sent to pilot school for flight training. Among them, some became aerial observers. Immediately after this cohort, another 40 cadets passed air observer training. The school was upgraded to a university in March 1935.

In 1933, a group of DH.82 Tiger Moth training aircraft at the IIAF in Tehran. (Author's archive)

In 1939, the Imperial Iranian Aeroclub was established on the order of Reza Shah. It had several branches, including one in Mashhad airport. IIAF DH.82s were used here for civilian student pilots; the first four of them are next to their instructor in November of that year. (Author's archive)

Polikarpov R-1s formed the backbone of the IIAF fleet's combat aircraft before the arrival of British-made Hawker Fury and Audax aircraft. Six featured in a military parade over Tehran on 22 February 1933. (Author's archive)

Reza Shah witnessed a formation flight of DH.82s and R-1s during a military parade over Tehran on 22 February 1933. Two DH.82s crashed and the investigation led to increased flight safety measures being implemented in the air force. (Author's archive)

One aircraft was equipped with landing skids for operations in snow. Polikarpov R-5, with s/n 36, is pictured at Ghaleh-Morghi airport in 1933. (Author's archive)

The Birth of the Imperial Iranian Air Force (IIAF)

When military aircraft were first used in the Iranian Armed Forces, their operator was the Aviation Department. On 29 October 1932, this department changed its name to Air Forces and its head was officially named as commander. On 23 May 1934, its name changed again, to Air Forces Command.

On 2 July 1933, on the order of Reza Shah, the Aviation Department was reorganised as the Imperial Iranian Air Forces. It now had a command, general staff and several subordinated units comprising 1st and 5th Regiments, a technical department and a maintenance centre (which later became an aircraft factory).

The 1st Regiment consisted of an Intelligence branch, a Photo-Reconnaissance branch, a Radio and Communication Group, a Transport branch and three operational squadrons comprising Liaison, Fighter and Observation-Bomber. The Liaison Squadron had four groups while the Observation-Bomber Squadron had five groups. The Photo-Reconnaissance branch photographed Iran's oil regions in Khuzestan and Qom for the first time in 1938.

The 5th Regiment was the pilot school, which had an Adjutant office, a pilot-training group, training, transport and maintenance branches, which in turn had aircraft maintenance hangars and the aircraft mechanic school (later university) within its organisation.

On 22 March 1935, the international name of the country was officially changed from Persia to Iran. The Imperial Iranian Air Forces (IIAF) had been known as Persian Air Force internationally, but this was later changed and the IIAF's name was recognised internationally as well.

According to Article 1 of Decree No. 2366, the title of the Iranian Military Aviation Organisation was changed from 'Air Forces' to 'Air Force' on 15 January 1937. This name did not change when the imperial government of Iran fell amid the rise of Islamists and the establishment of an Islamic autocratic government on 11 February 1979.

Hawker Audax and Fury in IIAF Service

With the change in structure and organisation of Iran's Military Aviation, a rapid expansion in the IIAF took place between 1933 and 1941. As a part of that, the Air Force was equipped with British-made fighter and bomber aircraft. In January and March 1933, 14 Hawker Audax two-seat reconnaissance bombers and 16 Hawker Fury single-seat fighters were ordered for the Air Forces.

By Iran's order, the Audaxes and Furies were fitted with Pratt & Whitney Hornet S2B1-G radial piston engines. Due to technical issues with these US-made engines in summer and autumn 1933, they were replaced with British engines. Another 14 Hawker Furies and 14 more Audaxes were ordered. Of these, four of the Furies were swapped for four more Audaxes in 1935. In that year, several more aircraft were ordered, bringing the total of Hawker Audax and Fury aircraft to 80.

In late February and early March 1934, the first British-made combat aircraft, all Hawker Audaxes, arrived by ship from Khorramshahr port. They were transported by heavy trucks to Ahvaz airport where they were assembled and test flown by Philip G Lucas, the British test pilot of the Hawker company, before the first four were ferried to Tehran on 8 April 1934. Deliveries continued until 9 May 1935. In total, almost 80 Hawker Audaxes and Furies were delivered. Junkers W33 transport aircraft were used by the IIAF to transfer pilots from Tehran to Ahvaz in order to ferry these aircraft.

These British combat aircraft were the first in the IIAF to have radio for communication. On 30 September 1934, an Iranian pilot made use of it for the first time during a flight from Tehran to Sultan Abbad and then Khorram Abad. The aircraft was an Audax with s/n 412, flown by 1st Lieutenant Gholam-Hossein Mostafavi (who retired with rank of Brigadier in the 1950s). The aircraft left Qaleh-Morghi at 08:00 local time and returned again at 17:40 the same day. In total, it logged three hours and 35 minutes, during which the pilot tested the radio system and allowed various air force radio stations to successfully test their gears.

In September 1934, the IIAF had its Furies re-engined with Bristol Mercury IV.S2s, starting with aircraft s/n 203, which was used as a testbed in the UK. Deliveries of re-engined aircraft took place between November 1934 and January 1935. In March 1935, 12 Audaxes were delivered with Hornet engines while the last 26 Audaxes, which were delivered in April 1935, had Bristol Pegasus IIM and IIM2 engines.

On 13 June 1934, Reza Shah Pahlavi travelled to Turkey at the invitation of Turkish President Mustafa Kemal Ataturk. During this trip, an IIAF officer, Captain Mahmoud-Mirza Khosravani, was among the military officers accompanying him. Five days after entering Turkey, Ataturk held a military parade in honour of Reza Shah. Reza Shah visited many cities, museums and air bases in Turkey, during which time the Turkish Air Force held display flights for him.

Reza Shah's trip to Turkey finished on 6 July 1934. His trip tightened and significantly improved the Iran–Turkey relationship so much so that the IIAF deployed five of its Hawker Audax light bombers to take part in the Turkish Republic Day parade on 29 October 1934.

The five Audaxes were piloted by the commander-in-chief of the IIAF and four other pilots. It was the first time in the IIAF's history that its aircraft exited national airspace. This deployment to Ankara, capital of Turkey, took place with refuelling stops at Tabriz and Eskişehir. Their deployment lasted until 9 November 1934, during which they logged a total 131 hours and 29 minutes of flight.

On 12 October 1936, one of the Hawker Audaxes crashed after its engine caught fire in the south-east of Ghaleh-Morghi airport. Its pilot, Warrant Officer Jaafar Darbandi, survived despite the fire burning his flight suit and a part of his parachute. He was the first pilot in the IIAF to jump to safety following the technical failure of an aircraft.

Doshan-Tappeh airfield, Tehran, 1938. A group of Hawker Fury fighters of the IIAF, ready for a formation flight during the official visit of Reza Shah. (Author's archive)

The 1934 wreckage of a Hawker Fury fighter that crashed during a ferry flight from Ahvaz to Tehran, discovered by the Air Force personnel and photographed by another aircraft. (Author's archive)

IIAF fighter-bomber pilots standing in front of a Hawker Audax equipped with a Pratt & Whitney Hornet S2B engine in 1936. (Author's archive)

Reza Shah meets a group of Turkish Air Force pilots who performed a display flight during his official visit to Turkey. (Author's archive)

A 1937 image of 402, a Hawker Audax equipped with Pratt & Whitney Hornet S2B engine and three-bladed propeller. (Author's archive)

A formation flight of five Hawker Audax bombers over Tehran during a parade. (Author's archive)

New Regiments in Ahvaz, Mashhad and Tabriz

On 15 November 1934, the IIAF engaged in a COIN in the Baluchistan region of Iran. The operation involved three Junkers W 33 transport aircraft and an old R-5 light bomber despite the availability of more modern Audaxes. The operation lasted until 24 March 1935. To make it easier for the IIAF to provide support for the COIN operations, new regiments were formed across Iran.

In January and February 1935, the IIAF received 20 DH.82As (c/n 3290 to 3309) and in May of the same year, ten more were delivered (c/n 3464 to 3473). Deliveries of the new aircraft allowed for the formation of new 3rd and 4th regiments in Ahvaz and Mashhad in south-west and north-east Iran, respectively. Each of these squadrons was equipped with seven Hawker Audax light bombers, seven Hawker Fury fighters and two de Havilland Tiger Moth training aircraft, which were used for pilot training and liaison duties.

On 14 March 1939, another regiment was established in Tabriz, north-west Iran, using mostly aircraft manufactured by the IIAF. These regiments were active until August 1941, when Iran was occupied by Allied forces during World War Two.

IIAF aircraft took part in their first Army exercise on 8 September 1935. The exercise, held by the IIGF's Tabriz Division, involved seven Audaxes piloted by four officers and three warrant officers. Six had observers, five of whom were Iranians, including Lieutenant Colonel Bevir who was foreign advisor of the IIAF. Following this deployment, the aircraft returned to Tehran on 21 September 1935.

Hawker Audax s/n 461 with Pegasus piston engine and two-bladed propeller at Doshan-Tappeh where it was partially built and assembled by the IIAF's Shahbaz aircraft factory in 1938. This aircraft was later sent to Ahvaz where it joined a new regiment. (Author's archive)

Hawker Fury s/n 208, which was stationed at Mashhad in 1938, during a photoshoot in the United Kingdom prior to delivery in 1934. (Author's archive)

According to the historical documents, on 1 October 1935, the IIAF had 51 Tiger Moths (33 DH.82s and 18 DH.82As), 29 Audaxes with Hornet engines, 28 Audaxes with Pegasus engines, 20 Furies with Mercury engines, ten R-5s with M-17 engines, two Junkers W 33s with L 5 engines, 11 R-1s with M-5 engines (five had their engines removed) and a Turkish-made Curtiss Fledgling equipped with a Wright Whirlwind engine.

Shahbaz Aircraft Factory

The Turkish government gifted a Turkish-built Curtiss Fledgling two-seat trainer to Reza Shah, which was flown to Iran by a Turkish Air Force officer in July 1934. Realising it was possible to build foreign-designed aircraft under licence, Reza Shah established an aircraft factory in Iran, with the intention of manufacturing such aircraft; the ultimate goal was to produce Iranian-designed aircraft in the future.

The IIAF was still in need of more aircraft, specifically Tiger Moths, Audax and Furies. To achieve this goal, the United Kingdom provided aircraft that required assembly and partial production. On the order of Reza Shah, the Iranian Ministry of War finalised deals in order to enable the IIAF to produce these three aircraft types domestically.

Swedish engineer N H Larsson was hired as head of the project to establish the aircraft factory, starting from August 1935. Construction was finished in 1936 and two British, one Italian and 41 Iranian mechanics and engineers were hired to work there. Among these, two Iranian officers and eight warrant officers were trained in the Hawker factory in the UK between November 1935 and February 1937.

Captain Walker from the UK and his assistant Frank Knight were hired by the IIAF as the factory managers and under their supervision, domestic production of the aircraft began with five DH.82A Tiger Moths, which were completed in October 1937. The aircraft were test flown successfully by the IIAF's chief test pilot, Captain Afkhami, in April 1938. The five Audaxes completed were delivered to various squadrons in May 1938.

With five more Audaxes and five other Tiger Moths under production, Reza Shah visited the factory on 21 June 1938 and requested it be named 'Shahbaz'. The second batch of Tiger Moths was complete in March 1939 and the Audaxes were complete in April and May. In total, 20 unassembled DH.82As with c/n 3940 to 3934, 82005 to 82009, 82047 to 82051 and 82092 to 82096 arrived at Khorramshahr port in May 1939 and were transported by train to Tehran, where they were assembled at Shahbaz aircraft factory from June 1939.

The Shahbaz-built Hawker Audax with s/n 461 during Reza Shah's visit to the factory and Doshan-Tappeh airport in 1938. (Author's archive)

The left side of the Hawker Audax with s/n 461 in Doshan-Tappeh airport in 1938. (Author's archive)

A line of Hawker Fury fighters including aircraft s/n 205 (nearest to the camera). Some were assembled and partially built in Shahbaz aircraft factory in 1938. (Author's archive)

A Hawker Audax and Fury on the final assembly line at the Shahbaz aircraft manufacturing factory in 1938. (Author's archive)

A partially manufactured de Havilland DH-82 Tiger Moth aircraft prior to shipping to Iran for assembly at Shahbaz aircraft factory. (Tiger Moth Society)

Reza Shah Pahlavi and his son Mohammad Reza (who became Shah in 1941) at Doshan-Tappeh airfield during the official visit to Shahbaz aircraft factory in 1938. (Author's archive)

Reza Shah Pahlavi and Mohammad Reza during their official visit to Shahbaz aircraft factory in 1938. (Author's archive)

Reza Shah and Mohammad Reza Shah in one of the hangars used by the Shahbaz aircraft factory. (Author's archive)

Reza Shah talks to Hawker Fury fighter pilots; their aircraft were built by the Shahbaz aircraft factory in 1938. His son, Mohammad Reza Pahlavi, stands behind him. (Author's archive)

Reza Shah next to a Shahbaz-built Hawker Audax light-bomber in Doshan-Tappeh in 1938. (Author's archive)

Hawker Hind in Service with the IIAF

Developed from the Hawker Audax, the Hawker Hind was a British light biplane bomber designed in 1933. The prototype flew on 12 September 1934, with the first examples entering service with the Royal Air Force in 1935. Iran became one of the first international customers, with a total of 35 ordered for the IIAF. Known as Hind Mk.I, they were equipped with a Bristol Mercury VIII radial piston engine. The Hind was superior to the Audax in some aspects, particularly in terms of manoeuvrability.

The first Iranian Hawker Hind, s/n 601, made its maiden flight in April 1938 following its delivery alongside 17 others by ship through Khorramshahr port. They were assembled in Ahvaz airport and, after test flights, were ferried to Tehran in November and December 1938. The remaining 17 aircraft were delivered by spring 1939. Domestic production of Hawker Hind Mk.Is started in December 1939 at the Shahbaz aircraft factory, with their deliveries starting in spring 1941.

The newly established airport of Mehrabad in west Tehran was selected as the new and permanent base of 5th (Training) Regiment, previously based at Doshan-Tappeh. In summer 1939, it had three Furies, eight Audaxes and all 35 of the newly delivered Hinds plus 51 Tiger Moths. The 1st Regiment at Ghaleh-Morghi had 15 Furies, 12 Audaxes and 20 Tiger Moths. The 2nd Regiment in Tabriz (previously in Ghaleh-Morghi) had ten Audaxes and five Tiger Moths. The 3rd Regiment in Mashhad had seven Audaxes and three Tiger Moths; while the 4th Regiment in Ahvaz had ten Audaxes and four Tiger Moths. Following completion of their training by pilots, the Hinds were delivered to the 1st Regiment, which passed its Audaxes to the 4th Regiment in Ahvaz.

In summer 1941, the IIAF had 55 Hind Mk.Is (one crashed), with s/n 601 to 655. They were operated by three regiments as light bombers, releasing most Audaxes for use by the 4th Regiment and other front-line regiments. Two Hawker Hinds plus nine Hawker Audaxes, all of which were dual control, were kept in service for pilot training.

In July 1940, one of the Hinds crashed due to engine failure at Tehran's railway station, north of Ghaleh-Morghi airport. Its pilot 1st Lieutenant Saeed Aazazi (later retired with rank of Brigadier) parachuted to safety.

This Hawker Hind with s/n 601 was the first of its kind to be built and delivered to the IIAF. It is in the UK before its maiden flight in April 1938. (Author's archive)

A Hawker Hind damaged in an incident in 1939. (Author's archive)

Airspeed Oxford, the First Twin Engine Aircraft of the IIAF

In 1924, Junkers Luftverkehr AG of Germany, one of the world's most successful airlines at that time, was contracted by the government of Persia to use the Junkers F 13 to carry passengers and parcels for the Persian Post and Telegraph. The Junkers F 13s and W 33s were widely used for these purposes until October 1931. The company's contract expired in January 1932 and flights ceased at the end of March. In 1933, three Junkers F 13s and five W 33s were transferred back to Germany.

To replace them, Reza Shah ordered new aircraft from the UK, in the shape of a pair of de Havilland DH.89 Dragon Rapide twin-engined passenger/transport aircraft, and two IIAF pilots, 1st Lieutenant Nafez and 1st Lieutenant Khan Baba-Ala, were selected to become their pilots. They were accompanied by the IIAF's chief test pilot Captain Abolfath Afkhami, who had acquired 33 hours and 10 minutes of flight experience with the aircraft in the UK.

The two DH.89s were delivered in September 1936 by ship through the port of Khorramshahr in the north-west of the Persian Gulf. They were ferried by road to Ahvaz airport where they were assembled, test flown and then ferried to Tehran. One more DH.89 had been ordered and was delivered a few days later. The three DH.89s received civil registration codes of EP-AAA, EP-AAB and EP-AAC in March 1938.

The AS.10 Oxford was a twin-engine monoplane aircraft developed and manufactured by Airspeed. It saw widespread use training British Commonwealth aircrews in navigation, radio-operating, bombing and gunnery roles throughout World War Two. In an attempt to equip its air force with multi-engine transport and bomber aircraft, the Iranian Ministry of War ordered three Oxford Mk.Is in autumn 1939.

EP-AAA and EP-AAB, two de Havilland DH.89 Dragon Rapide twin-engined passenger/transport aircraft operated by the PTT, are in an aircraft hangar at Ghaleh-Morghi, being viewed by Reza Shah and his son. They were the first of Iran's twin-engined aircraft which, despite being civilian, were maintained and operated by IIAF personnel. (Author's archive)

AS.10 Oxford Mk.I with s/n 803 (c/n P2002) is at Mehrabad airport in 1940. (Author's archive)

Oxfords with c/n P1984, P1993 and P2002 and s/n 801, 802 and 803 respectively, were delivered in March 1940 and assembled and test flown by Airspeed test pilot Brian Field before being ferried to Ghaleh Morghi airport. They became the first twin-engined aircraft operated by the IIAF.

On 29 June 1940, the 5th Bombardment Regiment of the IIAF was formed using the three Oxfords. Major Mahdi Sepah-Pour was appointed as commander of the unit. Oxfords could be armed with a Vickers K machine gun in a dorsal turret and could carry 16 practice bombs each of 11½lb (5.2kg) mounted externally, or four 100lb iron bombs. The aircraft had a bomb sight in the nose section, with an operator seated behind it.

Instead of being used to train future pilots of heavy bomber aircraft, Oxfords were used mostly for transport, liaison and even medical evacuation duties as the occupation of Iran during World War Two interrupted procurement of larger aircraft such as cargo, transporters and heavy bombers.

Hurricane Mk.I and Curtis H.75 Orders

On 21 December 1936, Brigadier Ahmad Nakhjavan, commander-in-chief of the IIAF, was appointed as deputy of the Iranian War Ministry. As a replacement for Nakhjavan, Colonel Ahmad Khosravani, commander of the 1st Regiment, was appointed as the IIAF's commander. On 20 January 1937, Lt Col Khalil Marjan was appointed as commander of the 1st Regiment.

In the years between 1932 and 1934, Swedish Air Force Colonel Elis Nordquist was the advisor and Chief of General Staff of the Aviation Department of the Persian Armed Forces. On 6 July 1934, Belgian Air Force Colonel Victor Bevir took the place of Nordquist. Finally on 21 April 1938, an Iranian officer, Major Mahmoud Khosravani, Bevir's deputy, was appointed Head of the General Staff of the IIAF.

On 9 February 1937, the first aircraft procurement commission or team was formed in the IIAF. It consisted of three officers named Captain Asadollah Dowlatshahi, 1st Lieutenant Karim Jenab and 2nd Lieutenant Ahmad Gerami.

In August 1940, another change in command structure of the IIAF took place with the formation of a Board of Directors or Commanders to run it. Brigadier General Ahmad Nakhjavan (Minister of War), Brigadier Abdollah Hedayat (head of the planning organisation of the Army), Brigadier Ahmad Khosravani (commander of the IIAF) and Colonel Khalil Marjan (commander of the 1st Regiment) were board members.

In the 1940s, the Supermarine Spitfire was selected by the Ministry of War as the future replacement for the Hawker Fury fighters in the IIAF. The Pahlavi government initiated an enquiry for procurement of 24 of them in November 1938, but the order was not confirmed by the UK and was later cancelled. As a replacement, Iran was offered the less advanced Hawker Hurricane Mk.I fighters; one of them, with c/n L2079, was shipped to Iran for a six-month test on 28 July 1939.

After assembly in Ahvaz, the Hurricane Mk.I was flown to Tehran by the factory test pilot Richard Reynell on 18 October. After successful tests in the hot and high climate of Iran, an order was placed for 30 Hurricanes with Merlin III engines. The British government agreed to provide half of this number, signing a contract for 15 Hurricane Mk.IIs with more advanced Merlin XX engines in December 1939. Iran requested that the British government provide material and parts to enable domestic production of the remaining 15 Hurricane Mk.IIs at Shahbaz aircraft factory. This request was approved. However, due to political reasons, the material was not delivered.

Deliveries of Hurricane Mk.IIs were delayed due to the time needed for design and development of a tropical filter for the engine's air intakes. Finally, the first Hurricane Mk.II, with c/n P3270 (s/n 252), was flown in the UK in May 1940. Fully painted in the IIAF's two-tone desert camouflage, the aircraft was seized by the British government and was not sent to Iran. An embargo was imposed on the country by the British government following its refusal to end its economic and political ties with Germany, resulting in confiscation of the Hurricane Mk.II and cancellation of the order in December 1940 leaving just one unarmed and inactive Hurricane in possession of the 5th Regiment at Ghaleh-Morghi airport.

Germany had a strong influence in Iran by 1939, which caused tensions with the UK in 1940. Prior to that, the Iranian government turned to the US for procurement of fighters. In November 1939, a military delegation consisting of Lt Col Mahmoud Khosravani, Major Shaltshi and Captain Abolfath Afkhami left Iran for the US to negotiate for the procurement of ten aircraft.

The US agreed to sell ten Curtiss H-75A-9 Hawk fighters. These, with c/n 15252 to 15261, were manufactured between 9 March and 8 April 1941, and received s/n 251 to 260. The first aircraft was shipped to Iran in March 1941 and the rest in May. A Curtiss test pilot, Sam Irwin, was sent to Iran to supervise their deliveries and test fly them following assembly in Ahvaz. After Iran's occupation in August 1941, the British forces captured all ten H-75A-9s, which were in disassembled status. Three were

One of the IIAF's Hurricane Mk.IICs, which was delivered after World War Two. This example, with s/n 2-15, lacks its guns, indicating its use as a pilot trainer. (Author's archive)

The third H-75A-9 of the IIAF (s/n 252) during a photoshoot in the United States in 1941. (Author's archive)

assembled and flown to Tehran. However, a decision was made to transfer all ten to the RAF. The RAF deployed them to India and used them as Mohawk IVs.

Various units of the RAF including two squadrons with Hurricane Mk.IICs were stationed in Tehran and to support them, the IIAF's Aircraft Factory in Dowshan-Tappeh was used by the RAF's No 138 Maintenance and Repair Unit to produce aircraft parts and perform heavy maintenance on its own aircraft while also maintaining the IIAF's aircraft.

Chapter 2
The IIAF During the Years of Iran's Occupation

The imperial government of Iran declared its neutrality at the beginning of World War Two, but this status did not prevent the governments of Great Britain and the Soviet Union from agreeing to invade Iran. The geographical position of Iran provided a strategic corridor for the transfer of American- and British-made weapons and military equipment to Russia, and it could facilitate the possibility of victory of the Red Army forces against Nazi Germany.

The Anglo-Soviet invasion of Iran, with code name Operation *Countenance*, started on 25 August 1941 and ended six days later after the Iranian government formally agreed to surrender on 31 August. A day earlier, it had accepted a ceasefire due to the Imperial Iranian Armed Forces being numerically and technologically outmatched by the British and Soviet armed forces. Prior to the surprise attack, two diplomatic notes were delivered to the Iranian government on 19 July and 17 August, requiring the Iranian government to expel German nationals. This constituted an attempt to justify the invasion politically.

The coalition forces invaded Iran from two fronts. From the Persian Gulf, the Royal Navy and Royal Australian Navy attacked the Imperial Iranian Navy bases and sank multiple ships, taking the lives of many Iranian Navy personnel, while British Commonwealth forces came by land and air from Iraq. The IIGF's armoured units lacked enough tanks and forces to oppose them in Khuzestan province. From the north, the 44th and 47th Armies of the Transcaucasian Front (General Dmitry Timofeyevich Kozlov) and the 53rd Army of the Central Asian Military District, supported by Red Air Armies, invaded Iran and occupied its northern provinces.

A few months before the Anglo-Soviet invasion, the IIAF had participated in the parade of the Imperial Iranian Armed Forces in Tehran on 22 February 1941. A flypast involving 104 Tiger Moths, Hinds, Furies and Audaxes was conducted by the IIAF. Despite the large number of aircraft in IIAF service, the sanctions imposed by Russia and the United Kingdom slightly reduced the number of active aircraft. Seventy per cent of its combat aircraft were active and airworthy at the time of the invasion, but were mostly not used as many of their air bases were targeted and the aircraft stationed there were damaged, some beyond repair.

At the time of the invasion, the IIAF's Commander in Chief was Brigadier Ahmad Khosravani; its General Staff was headed by Colonel Mahmoud Khosravani, while the commanders of the Air Regiments outside Tehran in the cities of Tabriz, Ahvaz and Mashhad were Colonel Gholam-Hossein Sheibani, Major Hedayat Gilanshah and Major Darab Jahansoozi, respectively.

Before the start of the attack, on the morning of 25 August 1941, six British twin-engine bombers flew from Iraq to Iran and dropped leaflets over the south-western and western cities of Iran. On the same day, the Soviet Red Army Air Force bombed the city of Tabriz and its outskirts. From that day until 27 August, the RAF and Soviet Red Army Air Force bombed various cities of Iran and air-dropped leaflets to break the morale of the Imperial Iranian Armed Forces and Iranian civilians.

On 25 August 1941, a command post was formed at the headquarters of the Imperial Iranian Joint Staff to command a counter-offensive operation but as the Pahlavi government accepted the ceasefire and surrendered, this command post was also dissolved on 30 August 1941.

2nd Regiment During the Anglo-Soviet Invasion

The IIGF's Tabriz Division was completely surprised on the morning of 25 August 1941 when the Soviet Red Army invaded Iran. After minor resistance at the border, troops from the division retreated. The IIGF's anti-aircraft units managed to shoot down a Soviet combat aircraft that day. The 2nd Regiment of the IIAF in Tabriz airport, meanwhile, carried out a reconnaissance sortie on 25 August 1941. The regiment kept its radio contact with Tabriz Division until 26 August 1941 when the communication line was cut.

After the IIAF's 2nd Regiment lost its contact with the IIGF's 2nd Division, the regiment's commander, Colonel Sheibani, informed the IIAF's HQ in Tehran about the situation. Subsequently, the Commander of the Air Force ordered him to evacuate to avoid losses to Soviet airstrikes. Subsequently between 01:00 and 03:00 local time on 27 August, pilots and mechanics prepared 15 aircraft for evacuation.

At around 03:00, the pilots, who mostly had no instrument flight rating and were not qualified to fly at night, left Tabriz with their overweight aircraft carrying heavy machine guns and spare parts. Among the 15 aircraft, 13 reached Zanjan airport for refuelling en route to Tehran. Two made emergency landings in the Talesh mountains due to engine failure. The aircraft were declared a loss, although their pilots and mechanics survived and travelled to Tehran by road.

4th Regiment During the Anglo-Soviet Invasion

A few months before the invasion, the 4th Regiment of the IIAF was on high alert due to riots in neighbouring Iraq. On 25 August 1941, after the Royal Navy occupied Khorramshahr and other regions in south-western Iran, the 4th Regiment remained uninformed about the situation as communication between the Imperial Iranian Navy (IIN) and the IIAF had been cut. Later that day, three RAF fighter aircraft attacked Ahvaz airport and used their cannon to strafe aircraft, hangars and personnel of the regiment.

As a result of the airstrike at Ahvaz airport, two non-commissioned officers, first sergeants Jamshid Sahmi and Seyed-Ali Tolouei and Second Sergeant Abbas Farzaneh-War lost their lives. Further British airstrikes were conducted against the airport, leaving only a handful of Iranian aircraft undamaged. These were used for reconnaissance missions on 26 and 27 August 1941. From 28 August 1941, activities of the regiment were stopped.

3rd Regiment During the Anglo-Soviet Invasion

On the morning of 25 August 1941, the 3rd Regiment in Mashhad airport did not encounter any enemy action; it took two more days for the Soviet Red Army to reach the border with Iran, On 27 August, several Soviet two-engined bombers flew over Mashhad and dropped leaflets. The commander of the 3rd Regiment, Major Darab Jahansouz, was tasked by the commander of the Mashhad's Division with transferring the regiment's aircraft to the reserve airport of the village in Toroq.

As the airport in Toroq was inadequate to protect the regiment's aircraft, the 3rd Regiment lost contact with the Mashhad Division's HQ, so the regiment's commander transferred his aircraft to Sabzevar, 170km west of Mashhad. All the airworthy aircraft were flown there, with one making an emergency landing en route at Kashmar due to bad weather conditions. Jahansooz, together with a Warrant Officer (technician) and with additional fuel and spare parts, left Sabzevar for Kashmar to assist its pilot Sgt Maj Reza Fallahi, the pilot of the grounded aircraft. However, Jahansooz's aircraft crashed and he and the mechanic on board lost their lives before reaching Kashmar.

1st Regiment During the Anglo-Soviet Invasion

On 25 August 1941, the 1st Regiment commanded by Lt Col Mohammad Moeini and its 1st Fighter Squadron, commanded by Major Mostafavi, were put on high alert as the Soviet and British forces invaded Iran from the north-west and south-west. Between 25 and 27 August 1941, the 1st Regiment conducted several daily air patrols over Tehran, protecting it from the threat of the Soviet fighters. At 13:00 local time on 30 August, when Major Mostafavi was giving a speech to personnel of the 1st Squadron, Captain Yadollah Amini, commander of the intelligence branch of the 1st Regiment, joined him and read a statement about the end of resistance and acceptance of a ceasefire by the Iranian government.

The squadron personnel, particularly the Hawker Fury and Hind fighter pilots, did not accept the order to surrender. They didn't believe the statement read to them by Captain Amini. Major Mostafavi asked him to bring a signed and stamped official order, as they didn't want to stop guarding and protecting the capital city. Mostafavi told Amini that if enemy aircraft approached Tehran, they would scramble their aircraft to intercept and shoot them down. Amini left but returned a few minutes later with an order signed by the 1st Regiment commander Lt Col Moeini, in which Mostafavi was ordered to end armed patrols over the city and avoid scrambling his aircraft.

Finally, Mostafavi notified his squadron's personnel that they were dismissed, as the unit was no longer tasked with confronting enemy aircraft. Not all of the officers and warrant officers accepted this order. Some decided to form a 'Resistance Group' within the IIAF in order to confront the enemy the next day. At night, they gathered in the office of Captain Ahmad Vasigh, commander of the 1st Group of the 1st Squadron, to assign tasks and missions to each member.

The resistance group was formed of eight people, and among them two were pilots. The group was headed by Captain Vasigh, while Captain Noorbehesht was appointed as armament officer; 1st Lieutenant Sajjadi was appointed as commander of the air base during the operation; 1st Lieutenant Khosravi was appointed as aircraft refuelling officer in the group; 2nd Lieutenant Vasegh was tasked to arrest Brigadier Ahmad Khosravani and Lieutenant Colonel Moeini; Lieutenant Colonel Kia was appointed as combat medic; Warrant Officer Ahmad Shoushtari was tasked to perform reconnaissance flights over the capital city; and Warrant Officer Mohammad Amir-Hamzeh was tasked to guard the aircraft.

In the morning of 30 August, just before the 1st Regiment commander gave a speech to personnel of his unit, Vasigh gathered a group of officers and warrant officers and invited them to resist as the Soviet Army paratroopers were going to jump over Tehran in an hour. Subsequently, Brigadier Khosravani and Lieutenant Colonel Moeini were arrested by Vasegh. Major Afkhami, who was present, attempted to use his handgun to confront Vasegh, who had pointed his gun at Khosravani. Before he removed his handgun from his holster, Vasegh shot him in the right hand. Afkhami grabbed the gun with his left hand, but Vasegh fired two more rounds at Afkhami, which hit him in the neck and cheek. Two members of the resistance group immediately put him in Brigadier Khosravani's vehicle and transferred him to hospital.

Warrant Officer Ahmad-Ali Vaziri, who had a personal grudge against Khosravani, was tasked to imprison him. In the process, Vaziri also severely beat Khosravani and imprisoned him in a storage room. The resistance group immediately took over the soldiers' house at the base and disarmed the soldiers. Those who joined the resistance group remained in the base while those who refused were dismissed. After Captain Vasigh heard that Ahmad-Ali Vaziri had imprisoned Khosravani in an unsuitable place, he transferred him to the IIAF's college and then dismissed Vaziri. Second Lieutenant Farshid was tasked with guarding Khosravani while he was imprisoned in the room of the dean of the college.

Several aircraft were armed and refuelled. Vasegh had prepared and printed thousands of leaflets to be dropped by the group over Tehran. Before dropping leaflets, two fighters (Hawker Hinds) piloted by Warrant Officers Shooshtari and Savad-Koohi were tasked with performing a reconnaissance mission over Tehran. These pilots subsequently noticed a pair of armoured personnel carriers headed toward Ghaleh-Morghi airport. The armoured carriers belonged to the 1st Armoured Battalion of the IIGF's Tehran's Division that had been sent to Ghaleh-Morghi to confront the resistance group. When the armoured carriers approached the gate of the air base, several members of the resistance group positioned themselves with rifles and machine guns to open fire but Vasigh ordered them to stand down.

When Captain Vasegh realised that the resistance group had no chance of surviving an attack of the IIGF's 1st Armoured Battalion, he ended the resistance, boarding an aircraft and left Tehran for Gilan in the north of Iran. Ebrahim Shooshtari, who was also loyal to Vasegh, followed in a second aircraft. Ahmad Khosravani was later released by his brother, Mahmoud Khosravani. He immediately arrested Vaziri and severely beat him.

After taking off from Ghaleh-Morghi, Shooshtari noticed gunfire from the armoured personnel carriers of the IIGF and his regiment's aircraft. He dived and strafed at one of them to temporarily halt their firing, but performed an emergency landing near Talesh after running out of fuel. He was wounded and treated at Tehran's central hospital while Captain Vasigh, the mastermind behind the failed resistance, managed to leave Iran successfully. After some months, Vasigh reached Germany, where he tried to enrol in the Luftwaffe but failed. Second Lieutenant Vasegh also left Iran while First Lieutenant Sajjadi, who remained in Qaleh-Morghi, was fired and later employed by the National Railway company.

When Captain Vasigh and Major Sergeant Shooshtari were flying over Tehran, the IIGF opened fire at them using machine guns. At the same time, an unarmed Soviet aircraft also appeared over Tehran on its way to Ghaleh-Morghi. Its pilot witnessed guns firing at his aircraft. He subsequently climbed and flew back to the USSR. After returning, he reported to Soviet authorities that despite the acceptance of a ceasefire by the Iranian government, his aircraft had been fired upon. Later that day, the Iranian government notified Moscow about the cause of the incident and prevented Tehran being bombed by the Soviets in response.

Hawker Fury fighter with s/n 203 during a photoshoot in the UK prior to delivery. This example was based at Ghaleh-Morghi airport on 25 August 1941. (Author's archive)

Above left: Warrant Officer Ebrahim Shooshtari, one of the IIAF pilots who refused to surrender to the Soviet Forces that occupied Tehran in August 1941. (Author's archive)

Above right: Warrant Officer Ebrahim Shooshtari in flight suit. (Author's archive)

The IIAF During the Early Years of Occupation

As a result of Iran's occupation by the Anglo-Soviet forces, the IIAF passed under direct command and control of the 1st Division of the IIGF, commanded by Major General Karim Boozarjomehri. Subsequently, all rules and regulations, as well as new directives of the IIAF, would be signed by Brigadier Ahmad Khosravani on behalf of Major General Boozarjomehri.

Iran's occupation interrupted pilot training in the IIAF as both fuel and oil were in short supply. In October 1941, out of 164 cadets on the ninth pilot training course, 87 left while 77 stayed. Of those who stayed, only nine officers and one warrant officer pursued pilot training. Unlike the pilot school, the technical school did not have its training sessions interrupted by the war.

On 23 September 1941, 124 warrant officers graduated as aircraft technicians, together with nine others as maintenance engineers. The technicians were absorbed by the flight squadrons, while the engineers were employed at the central maintenance centre of the IIAF at Doshan-Tappeh (formerly Shahbaz aircraft factory).

The decline in flying activities among the flight squadrons was demotivating for personnel. Many of the officers and warrant officers decided to leave the air force, while others were fired. On 6 November 1941, pilots, air observers and technicians had their wages increased in an attempt to improve motivation. In addition, some of those officers and warrant officers of the 2nd and 3rd Regiments who saved almost 30 aircraft from being bombed by the Soviets were promoted.

Reza Shah was sent to exile in Mauritius by the British government in an agreement that would bring his son, Mohammad Reza Shah Pahlavi, to power. He later died in the Parktown neighbourhood of Johannesburg, South Africa, on 26 July 1944.

In late 1941, the IIAF was again redesignated as an Aviation Department of the Iranian Armed Forces. Brigadier Ahmad Khosravani's position changed from commander to general manager. On 12 January 1942, however, he and the former head of the IIAF's General Staff, Colonel Mahmoud Khosravani were dismissed. As a replacement for Ahmad Khsoravani, a non-aviation expert, Brigadier Majid Firouz, was appointed as the head of the aviation department. His appointment brought anger and dissatisfaction among pilots and other personnel.

In February 1942, only 252 officers and warrant officers remained in the IIAF. While Brigadier Majid Firouz was the head of the Aviation Department, Colonel Sharafoldin Mirza-Qahramani and Colonel Gholam-Hossein Sheibani were his deputies. Despite a lack of airworthy aircraft, they deployed Audax light bombers to provide support for the IIGF's COIN operations against armed rioters in Kurdistan and Khorasan provinces in March 1942.

On 21 April 1942, following an agreement with the British government, Shahbaz aircraft factory was transferred to the RAF to be used for aircraft maintenance. In addition, Mehrabad and Qaleh-Morghi airports were transferred to the RAF and Soviet Red Army Air Force respectively. Subsequently, all

The RAF's Lockheed Hudson VI with s/n EW969, with its crew and passengers on 18 January 1943. (Author's archive)

of the aircraft of the 1st Regiment was transferred to Isfahan's airport in July 1942 while the 2nd Regiment (in Ahvaz) did the same in November 1942. The pilot school, which was in Mehrabad, also had all of its aircraft transferred to Isfahan, as the airport was transferred temporarily to the RAF during that year.

This centralised activities of what was left of the IIAF in Isfahan. With the exception of Isfahan, a handful of Hawker Hind fighters were deployed to Kermanshah, to be available for air interdiction and close air support missions during any possible COIN operations of the IIGF.

During the early years of Iran's occupation, the Aviation Department encountered management issues, as Brigadier Majid Firouz had no knowledge of aviation. On 27 August 1942, Brigadier Mohammad Nakhjavan (a pilot) was appointed as the head of Iran's Military Aviation, but was replaced by Colonel Sharafoldin Mirza Qahramani in December.

On 18 January 1943, Colonel Sharafoldin Qahramani and several other high-ranking officials in died when an RAF aircraft which was carrying them to Habbaniyah AB in Iraq crashed. The death of Colonel Qahramani once again created management issues. As Iran's Military Aviation had no chief or commander, a Technical Colonel named Mahna unofficially commanded the military aviation.

The RAF aircraft that crashed during its flight from Tehran to Habbaniya was a Lockheed Hudson VI with s/n EW969. It took six days for the wreckage to be found by members of the IIGF's cavalry in the Karaghan Dagh mountains near Nowbaran, 50km west of Saveh. All on board died, including five RAF crew. They were Leading Airman Stanley Ellis Gatley, Leading Airman Herbert Pilling Gilchrist, Group Captain Robert Ritchie Greenlaw, Leading Airman Rae Medina and Squadron Leader George Ewart Milnes.

In addition to Colonial Qahramani, the Hudson had six other passengers: three Iranian and three British officers. The Iranian officers were General Ibrahim Arfaa (commander of the IIGF's 2nd Central Division), Colonel Gholam-Hossein Shaibani (the IIAF's representative in the Joint Chiefs of Staff) and Colonel Abolfath Afkhami (commander of the IIAF's training centre). The British officers were Group Captain Greenlaw, British Air Attache in Tehran; Mr Stephen Childs, Director of Public Relations Bureau in Tehran; and Captain Yale, of British General Headquarters in Baghdad.

Military Operations During World War Two

Supported by foreign powers such as the Soviet Union, various tribes in Kurdistan, Khuzestan and the Fars regions of Iran started an insurgency against the Pahlavi government thinking its power was reduced due to Iran's occupation. The insurgency lasted just a few days after Iran's Military Aviation deployed its combat aircraft to support the IIGF's COIN operations against these insurgents, particularly in the Kurdistan region.

Hawker Hind fighter aircraft, the newest combat aircraft operated by the Iranian Armed Forces at that time, played an important role. On deployment in Kermanshah, they logged close air support and air interdiction sorties against the Soviet-backed Kurdish rebels.

In addition to Kermanshah, the Iranian Military Aviation deployed some of its combat aircraft to Shiraz airport from November 1942. They also played a similar role as they provided CAS for the IIGF units during their COIN operations against local tribal insurgents attempting to establish autonomy for themselves in Iran's Fars region.

Until mid-1943, all of the operational units of the Iranian Military Aviation Organisation or Department were based in Isfahan. Starting from August 1943, they were gradually transferred to Ghaleh-Morghi airport, leaving one squadron of aircraft behind. In addition to this squadron, the Iranian Military Aviation kept three aircraft groups deployed in Ahvaz, Kermanshah and Shiraz. In May 1944, Mohammad Reza Shah Pahlavi, the new king of Iran, paid a visit to the squadron.

Photographed during World War Two at Ghaleh-Morghi airport, a group of pilots and technicians stand next to a Hawker Audax. A Tiger Moth (s/n 161) is in the background. (Author's archive)

Avro Anson Mk.Is in IIAF Service

Iran's declaration of war on Germany on 9 September 1943 paved the way for the British Air Ministry to rebuild Iran's air power through the delivery of secondhand aircraft as replacements for those lost or damaged during and after the Anglo-Soviet invasion. The IIAF received an unknown number of Avro Anson Mk.I twin-engine, multi-role aircraft to be used for a variety of missions such as liaison, transport, multi-engine pilot training and heavy bombardment.

The RAF documents show deliveries of 48 of these aircraft to the IIAF in three batches between May 1944 and June 1945, while other sources indicate deliveries of just 14 Avro Anson Mk.Is and one Mk.XII, which was used by Mohammad Reza Shah Pahlavi and his family.

In June and July 1944, 12 pilot officers, three warrant officer pilots, three maintenance officers and four maintenance warrant officers were sent to Egypt to be trained on Avro Ansons. With their training complete, the aircraft were delivered in two batches, flying from Cairo to Iran over the Sinai desert, Syria and Iraq. They entered service with the 1st Heavy Bombardment Regiment in Isfahan.

In June 1944, the Aviation Department of the Imperial Iranian Armed Forces had 493 personnel, among them 80 were pilots and 60 were air observers. A few months after the end of World War Two, the Aviation Department had two flight regiments in Tehran and Isfahan, three independent aviation squadrons in Isfahan and three independent aviation groups, which were located in Kermanshah, Shiraz and Ahvaz.

One of the flight regiments was a Mixed Reconnaissance Regiment consisting of two reconnaissance squadrons, one independent fighter group and an Infantry Guards Company. The other regiment 'Bombardment' consisted of two squadrons served by Anson Mk.Is. This regiment was re-designated as a squadron due to the small number of its aircraft left in service.

On 22 August 1945, the IIGF launched a COIN operation in Bojnurd, north-east of Iran, home to the 8th 'Khorasan' division based in Mashhad. To provide close air support for the division, a unit consisting of five aircraft and a group of personnel was sent to Mashhad and officially formed its independent aviation group on 15 September 1945.

In March 1946, another independent Aviation Group was formed, this time in Hamedan's airport, using Hawker Hind fighters to carry out CAS missions in support of the IIGF during its COIN operations against the Soviet-backed Kurdish insurgents, a similar mission to the Kermanshah's independent Aviation Group.

An Avro Anson Mk.I with s/n 705 in Isfahan in 1946. (Author's archive)

Ghaleh-Morghi airport in 1945 with several USAAF C-47s present. (Pakzad, Iran's National Library)

Major General Donald H. Connolly of the US Army was appointed as commander of the Persian Gulf Command on 20 October 1942. The Shah of Iran is meeting him in Tehran in 1942. (Stetson Family Collection, Library of Congress)

Right: During his meeting with Major General Donald H. Connolly, the Shah visited a newly built US Army Air Force (USAAF) C-47 Skytrain. The Shah is seated in the aircraft. (Stetson Family Collection, Library of Congress)

Below: Major General Donald H. Connolly with Mohammed Reza Shah Pahlavi standing next to a USAAF C-47 Skytrain. (Stetson Family Collection, Library of Congress)

US aircraft stand ready to be picked up at Abadan field, a location that played an important role in the US military support for the Soviet Union during World War Two. Five principal types of aircraft were delivered to the USSR in May 1943, three of which are shown here. Of those delivered, about 20 per cent were Curtiss P-40s, 25 per cent Bell P-39s, 49 per cent Douglas A-20s, five per cent North American B-25s and one per cent North American AT-6s. (9th Air Force of the USAAF, Library of Congress)

The white star turns red. This symbolic paint application was made by an Iranian workman, substituting the Soviet for the American star on a Douglas A-20 before it was turned over in Abadan airport, Iran, in October 1943. (9th Air Force of the USAAF, Library of Congress)

A large number of Douglas A-20 attack aircraft prior to delivery to the USSR in Abadan airport on 17 February 1943. Iran played a strategic role in arming the Soviet Union during World War Two. In total, US deliveries to the USSR through Lend-Lease amounted to $11 billion in materials (equivalent to $148 billion in 2023) and included 11,400 aircraft comprising 4,719 Bell P-39 Airacobras, 3,414 Douglas A-20 Havocs and 2,397 Bell P-63 Kingcobras plus 1¾ million tons of food. (9th Air Force of the USAAF, Library of Congress)

Iran played a strategic role in the Allied victory in World War Two in Europe. Thousands of American- and British-made aircraft were delivered through Iran and most were assembled at Abadan airport. This image, taken on 26 August 1943, shows a complete P-39 fighter on the left and several crates of unassembled P-40s on the right, next to the aircraft hangar where they were assembled. This hangar was destroyed by Iran's Islamic regime in 2024 despite its historical value. (9th Air Force of the USAAF, Library of Congress)

Chapter 3
Rebirth of the IIAF after World War Two

On 30 June 1946, the aviation department was once again re-designated as the IIAF, a title lost due to the Anglo-Soviet invasion in 1941. This renaming took place on the direct order of the young Shah, who had just begun training to become an air force pilot.

Mohammad Reza Shah Pahlavi's pilot training began on 4 June 1946 in a DH.82A (s/n 192). After logging five hours and 25 minutes of training, he made his first solo flight in another DH.82A (s/n 166) on 25 June 1946.

Left: The Shah seated in DH.82A (s/n 166) after his first solo flight in Tehran, 25 June 1946. (Author's archive)

Below: On 17 October 1946, Mohammad Reza Shah Pahlavi received his pilot licence at Doshan-Tappeh airport upon completion of his pilot training. Major General Ahmad Nakhjavan, the first Iranian pilot, later appointed as commander of the IIAF and then head of the Iranian Civil Aviation Organisation, is presenting the licence to the Shah. (Author's archive)

On 1 September, he began training on different types of aircraft such as the Hawker Audax light bomber, Hawker Fury fighter, Avro Anson bomber and Beechcraft Model 18 transport aircraft (his private plane). On 17 October 1946, after logging 118 hours and 19 minutes of flight (across 271 sorties) in 132 days, the Shah graduated and received his wings.

Only one DH.82 Tiger Moth training aircraft remained by the 1950s. This was s/n 166, in which the Shah logged his solo flight on 25 June 1946. It was restored on the order of Brigadier Mansour Sattari, commander of the Islamic Republic of Iran Air Force (IRIAF), to be displayed in the now disbanded IRIAF museum in Tehran in 1992. Sattari and several other IRIAF commanders lost their lives when their aircraft, a Lockheed L-1329 Jet Star II, crashed near Isfahan on 5 January 1995. Just a few days later, the museum in Ghaleh-Morghi airport was closed and this aircraft was put in storage for almost 15 years. (Babak Taghvaee)

In January 2011, the DH.82 Tiger Moth (s/n 166) was disassembled and transferred by trailer from Ghaleh-Morghi airport to Doshan-Tappeh airport. (Babak Taghvaee)

DH.82 Tiger Moth (s/n 166) was later rebuilt but had its original fabric covering replaced with hard composite material. On the order of the Ideological and Political Department of the Islamic Republic of Iran Air Force, the aircraft's serial number was changed to remove the association of the Tiger Moth with the aircraft that the Shah used for his solo flight. Despite these changes, the museum authorities removed the DH.82 from its collection and disposed of it. (Babak Taghvaee)

Hurricane Mk.IICs in the IIAF Service

The sole Hurricane Mk.II that the IIAF received before the Anglo-Soviet occupation was absorbed by the RAF at Mehrabad airport, Tehran. It had no guns and hadn't been flown for months. Between December 1942 and March 1943, the RAF's No. 74 Squadron was stationed at Mehrabad airport, where British forces operated Hurricane I/IIs to protect the city from any possible attack by the Luftwaffe.

By May 1943, No. 74 Squadron had left Tehran, leaving behind ten Hurricanes (mostly Mk.IIBs). After a year, these were delivered to the Imperial Iranian Armed Forces. It is not known whether the sole Hurricane Mk.I that the IIAF received before the war (L2079) was among them.

According to the IIAF's historical documents, there is no indication that the ten ex-74 Squadron Hurricanes were put into service in 1944. The Air Force officially began operating Hurricanes when brand new Mk.IICs ordered by the Iranian Ministry of War (the old order for 18 Mk.IIs having been changed to Mk.IICs in 1945) were delivered.

In total, 16 Mk.IICs were built by Hawker for Iran, with the first group being delivered by ship through the port of Khorramshahr in summer 1946. After assembly in Ahvaz, they were flown to Ghaleh-Morghi airport. In addition to these aircraft, at Iran's request, two Hurricane Mk.IIBs were converted to a special two-seater training aircraft, equipped with the same wing as the Mk.IIC. They were named Mk.T.IIC, with 'T' suffix for 'Training'.

The first of the two-seater Mk.T.IICs, (KZ232) first flew on 27 September 1946. It had a maximum take-off weight of 8,140lb (3,692kg) without armament and could reach a maximum speed of 320mph

A handful of Hawker Fury fighters were left airworthy in 1946, but they made way for more modern Hawker Hurricane Mk.IIC fighters that year. This Fury with s/n 208 was one of them. (Author's archive)

Hurricane Mk.IIC with s/n 2-13 during a photoshoot in England prior to its delivery to the IIAF in 1946. (Author's archive)

(515km/h) at 21,500ft (6.55km). This was slightly different to the 8,710lb (3,950kg) maximum take-off weight of the Hurricane Mk.IIC (fully armed), which had a maximum speed of 340mph (547km/h) at 21,500ft (6.55km).

The Iranian Hurricane Mk.IICs were equipped with Rolls-Royce Merlin XX liquid-cooled V-12 piston engines producing a maximum 1,185hp (883kW) power at 21,000ft (6,400m). This enabled them to carry four 20mm (0.79in) Hispano Mk.II cannons in their wings and two 250 or 500lb (110 or 230kg) bombs under the wings, similar to examples that served with the RAF.

2-31 was one of two Mk.T.IIC two-seater training aircraft that the IIAF received in 1946. Here it is in England with an open canopy for the aft cabin. This was later replaced with a streamlined closed canopy. (Author's archive)

Another image of the Hurricane Mk.T.IIC two-seater training aircraft 2-31 with open aft canopy in England in 1946. (Author's archive)

The IIAF's Role in the Liberation of Azerbaijan Province

Following the end of World War Two, the British and American armies left Iran on 2 March 1946. However, the Soviet forces remained. This decision contravened the agreement reached between the heads of the Allies during the Tehran Conference on 1 December 1943. The presence of the Red Army in the province of Azerbaijan, in the north-west of Iran, guaranteed Moscow's access to Iran's oil resources in the Caspian Sea.

In order to legitimise the presence of the Red Army in Iran, the Soviet Union established the Democratic Party of Azerbaijan in October 1945. This sect declared the independence of Azerbaijan from Iran on 12 December 1945, and shortly after, the People's Republic of Mahabad was established in the west of Azerbaijan province by the Kurdistan Democratic Party with the support of Moscow. Puppet governments were created with the purpose of legitimising the presence of the Soviet military in the north-west of Iran.

After the occupation of Azerbaijan, the government of the Soviet Union tried to impose its orders on the young Shah and force him to accept the separation of Azerbaijan from Iran, taking advantage of the weakness of the country's armed forces, caused by several years of occupation. In response, Mohammad Reza Shah Pahlavi said, 'Even if my hands are cut off, I will not sign the document of separation of Azerbaijan.'

The imperial government of Iran now filed a complaint against the Soviet Union to the United Nations. It wasn't long before the Pahlavi government received support from the United States and its president, Harry Truman, in the political conflict with the Soviet Union.

The Soviet Union did not wish to confront the USA so withdrew the Red Army from the occupied areas of Iran. The withdrawal provided an opportunity for the imperial army to defeat the military branch of the Azerbaijan Democratic Sect, known as Fedayans, and liberate the occupied areas on 12 December 1946. The liberation of Mianeh followed the victory of the imperial army in a battle with the loyalists of the Azerbaijan Democratic Sect in Ghaflankooh Valley on 11 December 1946.

The IIAF carried out multiple photo-reconnaissance flights using Avro Anson Mk.Is 17 days before the start of the operation. The first such mission was carried out on 22 November 1946 by an aircraft with s/n 707, piloted by the Shah. In seven hours, the aircraft flew over the Qazvin, Kermanshah and Kurdistan regions.

On 25 November 1946, a similar mission was carried out by an Anson with s/n 704, travelling from Qaleh Morghi airport to Azerbaijan. After three hours of reconnaissance, it landed in Zanjan, returning to Qaleh Morghi on the same day. Another Anson (s/n 713) flew a similar mission under the control of the Shah on 29 November 1946.

On 11 December 1946, the Hurricane Mk.IICs played an important role in bombing strongholds of the separatists before the IIGF retook control. Five days earlier, on 6 December, these aircraft were forward deployed to Zanjan's airport while a chartered Bristol Type 170 Mk.32 Superfreighter heavy transport aircraft was used to airlift ground crew, ground equipment and additional weapons and ammunition from Tehran to Zanjan.

On 6 December 1946, the Chief of General Staff of the Imperial Iranian Army, Lieutenant General Haj Ali Razmara together with Mohammad Reza Shah Pahlavi, flew from Tehran to Zanjan and paid a visit to the combat aircraft, which were ready to carry out combat missions. On the same day, Mohammad Reza Shah Pahlavi flew with one of the Hurricanes over Azerbaijan and was pictured next to the aircraft after the mission. The image was published in newspapers the next day and broke the morale of the Communist separatists.

Hurricanes were not the only combat aircraft of the IIAF that participated in the operation. On 26 November 1946, the IIAF deployed several combat aircraft to Zanjan; a Hawker Hind with s/n 609 was the first to land. Avro Anson Mk.Is were used to carry ground crew, equipment and spare parts while

The IIAF lacked any heavy transport aircraft to carry personnel and equipment when the operation for liberation of Iran's Azerbaijan province was launched by the IIGF in 1946. The Anglo-Iranian Oil company had Bristol 170 Freighter Mk.21 aircraft provided them for this purpose. G-AHJD was one of them, and was in use by the company until 1951. (Author's archive)

A Bristol 170 freighter Mk.21 from the Anglo-Iranian Oil company with G-AHJD registration. (Author's archive)

weapons, ammunition, fuel and a truck with a photographic laboratory for printing images taken by reconnaissance aircraft were sent from Tehran to Zanjan.

A day after the liberation of Azerbaijan, the IIGF's forces entered the city of Tabriz on 12 December 1946. On the same day, several of the IIAF's combat aircraft landed in the city's airport, which the separatists had used. The small air force of the separatists was formed of only two hijacked aircraft, a civilian Rearwin Cloudster Model 8125 training aircraft from the Imperial Iranian Aero Club and a Hawker Hind fighter from the IIAF. They had only two pilots and no mechanics or weapons to use against the advancing government forces.

Iran's First International Aviation Competition

In 1951, on the order of Mohammad Reza Shah Pahlavi, the first international aviation competition in Iran was held, involving both civil and military aircraft piloted by Iranian and international pilots. The fighter aircraft in Qaleh-Morghi participated in the competition, which included Hurricane Mk.IICs. Some of the aircraft, such as Avro Ansons, had multiple crew members, while the fighters had single crew on board, thus their pilots could gain better scores in perfect piloting and navigation.

The competition started at Mehrabad airport on 31 May 1951. Several pilots and navigators competed with each other on board 14 aircraft for perfect navigation, formation-flight landing with a failed engine, and so on. They all landed in Isfahan the same day and then continued to Shiraz the next day. A few days later, when the aircraft were returning to Tehran via Isfahan, their pilots encountered a storm that resulted in the crash of two Hawker Hurricanes. The pilot of the first aircraft, Captain Eftekhari, lost his life, while the pilot of the second aircraft survived.

The competition concluded on 3 June 1951. That day, a ceremony was held in Mehrabad airport, during which the Shah attended to present the winners with their prizes. He gave a speech and asked for one minute's silence as a mark of respect for Captain Eftekhari.

Deliveries of 244 Aircraft by the US Under MDAP

Other than several Hurricane Mk.IICs, DH.82A Tiger Moths and Avro Anson Mk.1s, the IIAF did not receive any other aircraft from the United Kingdom during and after the end of World War Two. Starting in 1948, the Pahlavi government improved its relationship with the United States, its key strategic ally in the Middle East. Under a Mutual Defense Assistance Program (MDAP), the IIAF received a total of 244 piston-engined aircraft over a period of eight years, including 111 examples in 1949.

Out of those initial 111 aircraft delivered by the United States under MDAP, seven were Douglas C-47A Skytrain transport aircraft, which were used to replace a number of Avro Anson Mk.Is that had been damaged or lost in incidents and accidents; 37 L-4H Cub liaison aircraft, which relieved some of the DH.82As from liaison duties; 15 PT-13D light training aircraft, which replaced some of the old DH.82s; seven armed AT-6Ds capable of firing unguided rockets and bombs as replacements for some of the Audax light-bombers; and 45 F-47D (previously known as P-47D) fighters as replacements for part of the Hawker Fury and Hind fleet.

Up to 1957, the IIAF received 125 more American-made piston-engined aircraft, which had been used by the United States Army Air Forces (USAAF) during World War Two. They were seven C-47A/Bs, 21 L-4Hs, ten PT-13Ds, 16 T-6Ds and AT-6Ds, 52 LT-6Gs (delivered between 1955 and 1957 as replacement for the L-4Hs), four RLTG-6G light reconnaissance aircraft, and 15 more F-47Ds (delivered in 1950 as a source of spare parts.

LT-6Gs were multi-purpose aircraft. They were used for liaison duties and training purposes as well as close air support as they could be armed in a similar way to AT-6Ds. They could carry unguided bombs and rockets and two gun pods under their wings. To relieve them from liaison duties, the IIAF also received eight brand new de Havilland Canada DHC-2 Beaver Mk.I (L-20B) liaison aircraft from the USAF under the MDAP in 1954. The aircraft were initially used by IIAF generals but later were used to transport base commanders as well.

In 1947, the IIAF resumed training new air and ground crews in its training centres. In a period of five years, the training centres were expanded to improve the quality and level of officer and warrant-officer training with the help of the USAF advisors from the US Military Assistance Advisory Group (MAAG). A number of the IIAF's pilots were sent to West Germany to undertake training on the new aircraft, such as F-47Ds, in 1948. At the same time, the USAF trained the IIAF's maintenance engineers and mechanics in France and the United States. The new training aircraft delivered by the US under the MDAP were used to train hundreds of new pilots via the IIAF's instructor pilots, who had been trained by the USAF in West Germany.

The F-47Ds, which were delivered in 1949, entered service with the 2nd Fighter Regiment at Ghaleh-Morghi airport while the 1st Fighter Regiment operated Hurricane Mk.IIC and Mk.TIICs. The older fighters and light bombers were given to Number 1 and 2 Reconnaissance Squadrons as well as the units in Tabriz, Mashhad, Ahvaz, Shiraz and Isfahan. There were 25 Hawker Hinds, seven Hawker Audaxes, 12 DH.82 Tiger Moths and two Avro Ansons.

In addition, the pilot school also operated a large number of older fighters and light bombers as well as its training aircraft. In 1950, the school had 17 DH.82s, five Ansons and four Audaxes. The school also had a large number of grounded aircraft such as 17 DH.82s, 15 Audaxes, seven Hinds, five Ansons, three F-47Ds, two Oxfords and two Hurricane Mk.I/IIs. These grounded aircraft were replaced with 15 PT-13Ds and seven AT-6Ds. With deliveries of more PT-13Ds and AT-6Ds, the active and airworthy Hinds, Audaxes and Tiger Moths were retired. The PT-13Ds were retired and replaced with LT-6Gs from summer 1954.

Prior to their retirement, the PT-13Ds were equipped with spraying equipment and were used to spray land affected by locust infestations in the early 1950s. Between 21 March and 5 August 1952, six PT-13Ds were dedicated to air-spraying operations in support of the Iranian Ministry of Agriculture to combat locusts in Fars and Khuzestan provinces. This task was later undertaken by C-47A/Bs, which were equipped with a domestically designed aerial-spraying system in 1956.

The table shows delivery years of 244 piston-engined aircraft to the IIAF by the USAF under the MDAP between 1949 and 1957:

Aircraft type	Role	1949	1950	1951	1952	1953	1954	1955	1956	1957	Total
Douglas C-47A/B Skytrain	Transport aircraft	7	-	-	-	-	-	-	2	5	14
Piper L-4H Cub	Liaison aircraft	37	6	-	5	5	5	-	-	-	58
De Havilland Canada L-20B Beaver	Liaison aircraft	-	-	-	-	-	8	-	-	-	8
Stearman PT-13D Kaydet	Primary Trainer	15	2	2	2	2	2	-	-	-	25
North American T-6D/AT-6D Texan	Advanced Trainer	7	-	6	-	5	5	-	-	-	23
North American LT-6G Texan	Liaison aircraft	-	-	-	-	-	-	36	-	16	52
North American RLT-6G Texan	Reconnaissance aircraft	-	-	-	-	-	-	-	4	-	4
Republic F-47D Thunderbolt	Fighter aircraft	45	15	-	-	-	-	-	-	-	60

The AT-6Ds and LT-6Gs also played an important role in the COIN operation of the IIGF's 6th 'Fars' Division against the Javanroudi between 1952 and 1954. They mostly belonged to the aviation squadrons of the IIAF in Ahvaz. In addition, eight other aircraft from Qaleh-Morghi were deployed to Shiraz and were used on Hengam and Qeshm Islands in the Straits of Hormuz. On 7 April 1956, a pilot and an observer onboard an LT-6G died after their aircraft crashed while returning from Ahvaz to Isfahan after participating in the operation. In those years, the LT-6Gs as well as Avro Ansons were used for maritime patrol over the coasts of Persian Gulf and Hormuz Strait to monitor commercial ships and military vessels sailing in these waters.

This is one of 15 Stearman PT-13D Kaydet training aircraft that the IIAF received in 1949 for pilot training but was later used mostly for aerial spraying. This example had s/n 1-002. (Author's archive)

During World War Two, the RAF delivered some of its surplus DH.82 Tiger Moth training aircraft to the IIAF. They were mostly distinguishable from those delivered before the war as their engine cowlings were partially painted yellow or orange. (Author's archive)

During the 1950s, LT-6Gs delivered by the US to the IIAF under the Military Aid Program formed the backbone of the light attack aircraft fleet in the air force. Here, one of them with s/n 6-39 is at Shiraz airport in 1958. An American pilot stands in the middle while the pilot of the LT-6G stands on his left and the observer on his right. (Author's archive)

One of a few surviving T-6s in Iran. This example was an AT-6D pure attack variant that was displayed in the now closed museum of the IRIAF in Ghaleh-Morghi airport. It was transported to Doshan-Tappeh airport in January 2011, just a few weeks before the official closure of Ghaleh-Morghi airport and its conversion to a park. (Babak Taghvaee)

Only one L-4H was kept from being scrapped in the late 1950s. It is at Doshan-Tappeh, prior to being displayed in the IIAF museum there in 1972. (Author's archive)

The sole surviving Piper L-4H of the IRIAF, stored between 1995 and 2011, was severely damaged after a storm destroyed its hangar roof in 2005. Large piles of wood fell on this aircraft and damaged its structure. (Babak Taghvaee)

The sole surviving Piper L-4H liaison aircraft prior to its transportation to a larger museum at Doshan-Tappeh airport. It was restored and put on display in the larger museum, which never opened. (Babak Taghvaee)

Dar Cheshm-e Baad (*In the Wind's Eye*) was a propaganda show that state television produced between 2003 and 2008 to discredit the Pahlavi government, particularly Reza Shah. During the filming of a scene related to the events of 30 August 1941 in Ghaleh-Morghi airport, the door of the oldest aircraft hangar of Iran in that airport was left open and an aircraft was kept hanging from spars of its roof, resulting in its collapse in heavy wind. The damage to hangar was never repaired. (Babak Taghvaee)

This L-20B Beaver liaison aircraft (s/n 6-211) in olive-drab colour was used by Lieutenant General Nader Jahanbani, deputy commander of the IIAF in the 1970s. (Author's archive)

One of only four surviving L-20B Beaver liaison aircraft left in Iran can be seen here. Its colour scheme was changed from olive-drab to silver for use in the *Dar Cheshm-e Baad* (*In the Wind's Eye*) television show in 2006. It was later restored and repainted olive drab after it was transferred to Doshan-Tappeh in 2012. (Babak Taghvaee)

Mohammad Reza Shah Pahlavi stands next to an L-20B Beaver, which he piloted for short distances during military exercises near Tehran in the late 1950s and early 1960s. (Author's archive)

The Persian Thunderbolts: The P-47Ds of the IIAF

For air-to-air interception, the IIAF had Hurricane Mk.IICs, but for ground attack and other air-to-ground missions, Hawker Hinds and Audaxes were old and could not carry many bombs or additional weapons such as rockets or gun pods. To meet the needs of its 2nd Fighter Regiment for a more capable fighter-bomber, the P-47D or F-47D was offered by the USAF to the IIAF.

This aircraft was a successful high-altitude fighter, serving as the foremost USAAF fighter-bomber in the ground-attack role during World War Two. It had eight 0.50in-calibre machine guns, and could carry 5in rockets or a bomb load of 2,500lb (1,100kg). With a maximum take-off weight of eight tons, it was one of the heaviest fighters of its era.

In total, 12,558 out of 15,683 P-47 Thunderbolt fighters manufactured between 1941 and 1945 were 'D' models. Nicknamed as 'Jug' due to its appearance, the aircraft was the first P-47 model to have a bubble canopy. It was the first version of the aircraft to undergo large-scale production following an order placed by the USAAF for 850 of them on 14 October 1941 (later increased to 12,500).

Compared to older models, particularly the P-47C, the P-47D enjoyed many improvements, including changes in the turbo-supercharger exhaust system, additional cowl flaps for improving engine-cooling airflow and more importantly, extensive armour around the cockpit to protect the pilot. Hardpoints for a belly tank or a 500lb bomb were added to the P-47D-5-RE (D-11-RA) and later blocks including the P-47D-30-RA, which the IIAF received from the USAF. The P-47D-30-RA could carry zero-length launching stubs for a total of 16 5in HVAR rockets under their wings.

Like the P-47D-27-RE production lot, these P-47D-30-RAs had dorsal fins fitted to the head of the vertical stabiliser, which prevented directional instability due to the addition of the bubble canopy and removal of the aft keel area.

The P-47D-30-RAs received by the IIAF were all surplus USAF aircraft that had mostly not flown since 1946, as their units were disbanded or had upgraded with jet aircraft such as F-84E and F-84G fighter-bombers. At-least 25 of the 60 Thunderbolts were previously operated by the USAAF's fighter squadrons (FSs) from 9th Air Force (9th AF) in Belgium, France and Germany until 1945 or 1946.

The ex-9th AF Thunderbolts had served with the 22nd, 365th, 366th, 377th, 386th, 387th, 391st, 392nd, 395th, 396th, 397th, 405th, 406th, 508th, 509th, 512th and 513th Fighter Squadrons. At least three other examples delivered to the IIAF had served with the 83rd FS, 78th FG from the USAAF's 8th AF in Duxford, UK, while 15 others had been used by the 12th AF. The 12th AF units that operated the 15 aircraft were 86th, 313th, 314th, 522nd, 524th, 526th and 527th FS.

Serviceability of the F-47Ds

Out of the 45 F-47Ds that the IIAF put into operational use, six were lost in incidents and accidents up to 1952. According to USAF statistics, only 11 of the 45 were active as of 30 September 1952. Most of the 28 inactive examples were grounded due to technical issues surrounding their Pratt & Whitney R-2800-59 Double Wasp 18-cylinder air-cooled radial engines.

Four F-47D fighters ready for flight with a group of the air force personnel, including maintenance officers, standing in front of them at Doshan-Tappeh airport in 1952. (Author's archive)

F-47D-30-RA Thunderbolt with s/n 2-85. This specific airframe served with the USAAF's 9th Air Force during World War Two and after that with s/n 44-33126 (c/n 4087). The 9th Air Force's historical documents show it was operated by the 395th Fighter Squadron 'Panzer Dusters', 368th Fighter Group between 1944 and 1946. The squadron was deactivated at Straubing, Germany, on 20 August 1946 with all of its P-47Ds (later F-47Ds) being declared surplus. (Author's archive)

Later in December 1952, the number of airworthy F-47Ds was increased to 15 but to keep this number steady, the IIAF had to cannibalise some of its grounded examples. Combined with more accidents and incidents, this reduced the number of F-47Ds in service to just 28 on 30 June 1953, with only six of these still airworthy!

Another F-47D crashed, reducing the total to 27 with only seven of being airworthy on 30 September 1953. In just three months, however, the IIAF received a large number of spare parts and its mechanics and maintenance engineers managed to restore all 20 grounded F-47Ds to service. From December 1953 until 1956, when the type was withdrawn from service in the IIAF, the Fighter Bomber Wing of the air force in Ghaleh-Morghi had no grounded F-47Ds.

In 1954 and 1955, the rate of technical failures and number of incidents and accidents increased significantly. Many damaged aircraft were not repaired and were cannibalised for their parts instead. This reduced the total number of F-47Ds to just 20 on 30 September 1954. This number dwindled further, to just 17 on 31 December 1954. To improve the situation, some of the F-47Ds which had been withdrawn from operational use and allocated for spare parts were repaired and restored, increasing the total number of these aircraft in IIAF service to 26 as of 30 June 1955.

Some of the first F-47D pilots of the IIAF's 2nd Fighter Regiment in Ghaleh-Morghi. The pilot seated in the middle of the front row is Mohammad Khatami, commander of the regiment at that time and later commander-in-chief of the IIAF. The second pilot standing from the left is Nader Jahanbani, founder of the Golden Crown aerobatic display team of the IIAF and later, deputy commander of the air force. (Author's archive)

In 1956, the IIAF received its first jet aircraft, two Lockheed T-33A Shooting Star trainers. Four more T-33As, together with 54 more Republic F-84G Thunderjet fighter-bombers, were delivered under the MDAP in 1957. Another 21 of these fighter-bombers were delivered in 1958, which allowed complete retirement of the F-47Ds from IIAF service in early 1957. The F-47Ds used for their spare parts were scrapped at Doshan-Tappeh airport, while around 25 airworthy F-47Ds were returned to the US at the end of 1956. Among these, several aircraft ended up in the Turkish Air Force and were used until the end of 1957. No Thunderbolts were left in Iran.

The Persian Dakotas: The C-47A/Bs of the IIAF

Between 1949 and 1972, 30 C-47A Dakota transport aircraft and an EC-47D Electronics Calibration aircraft were operated by the Imperial Iranian Air Force (IIAF). They were used on a variety of missions from tactical transport to disaster relief. They flew across the Middle East, Africa, Asia and Europe and were also used in counter-insurgency operations inside Iran. After the delivery of four C-130Bs and eight C-130Es to the IIAF between 1964 and 1969, they were gradually sidelined from their major transport roles and then, upon delivery of the Fokker F27-400M/600s and more C-130E/Hs in 1972, were fully retired from service.

The first seven C-47As that the IIAF received in 1949 had c/n 10237, 19442, 19629, 19669, 19682, 20130 and 20168. They had served the USAAF with s/n 42-24375, 42-100979, 43-15163, 43-15203,

Before delivery of the C-47A/Bs by the US government under the Military Air Programs, the IIAF relied on its Avro Anson twin-engined training/bomber aircraft for transport missions. They were used by Mohammad Reza Shah Pahlavi until he flew his own aircraft in 1947. This image taken in the late 1940s shows the Shah exiting an Avro Anson aircraft. (Author's archive)

43-15216, 43-15664 and 43-15702, respectively. In service with the IIAF, they received s/n 5-01 to 5-07. IIAF formed its Transport Squadron with these seven aircraft in May 1949. One of the C-47As, 5-01 with c/n 10237, was later transferred to Iranian National Airlines in 1949 and received civil registration EP-ADG. Together with a number of other C-47s in use by the airline, it was operated by civilian and air force crew.

On 11 September 1949, 5-02 (c/n 19442) was used to transfer two IIAF officers to Paris, France, for an Air Staff training course. They were Air Watch Colonel Mohammad Kamal and Pilot Lieutenant Colonel Noor-Behesht. The aircraft was under control of Lieutenant Colonel Mansour Rahmani (navigator), Major Ali-Mohammad Khademi (pilot who later became a Major General and the founder of modern Iran Air in the late 1960s) and Major Hadi Khosravani (co-pilot). 5-02 reached Paris on 13 September.

5-02's flight to Paris was the first long-range flight of an IIAF aircraft to Europe. After that flight, Major Khademi was promoted to Lieutenant Colonel on 23 September 1949.

The second long-range flight of an IIAF's C-47A took place on 12 September 1949 when another C-47A left Mehrabad Airport for Nuremberg, West Germany. Passengers included Staff Colonel Nouri Alaei, commander of IIAF's Staff, and Colonel Hedayat Gilanshah, who later became commander in chief of the IIAF) then commander of the Fighter Wing of IIAF. They attended a military exercise and later briefed IIAF's high-ranking personnel, commanders and pilots about it.

Similar flights of Dakotas to West Germany transporting high-ranking officers to witness a NATO exercise were repeated six years later. On 18 June 1955, a C-47A piloted by 1st Lieutenant Abdollah Kohan and 1st Lieutenant Habollah Moeen-Zand, under command of Brigadier General Mostafavi, was used to transfer a number of high-ranking officers of Iranian Armed Forces with different expertise headed by Brigadier General Qodratollah Khazaei. The representative of the IIAF in the exercise was Lieutenant Colonel Manuchehr Pajuh-Afsar, who returned to Tehran together with other Iranian officers on 14 July 1955.

In April 1947, Trans World Airlines displayed a B-17G Superfortress heavy bomber equipped with a VIP cabin to the Shah. This former military aircraft was used by the USAAF as 44-85728, and was maintained and operated by the IIAF under the civil registration EP-HIM. It was flown by the Shah for domestic and International journeys between 1947 and 1949 when its place was taken by a smaller and more economical to maintain aircraft, a Beechcraft D.18S. (Author's archive)

On 29 January 1951, IIAF established the Military Air Transport Service with six C-47A Skytrains (s/n 5-02, 5-03, 5-04, 5-05, 5-06 and 5-07). A ceremony was held in Ghaleh-Morghi airport on that day. (Author's archive)

The first C-47A of the IIAF, s/n 5-01, was delivered to the newly established Iranian National Airlines (later Iran Air) in 1949 and received registration code EP-ADG. (Author's archive)

This image shows the formation flight of four C-47As over Tehran during a military parade in 1950. They had an overall olive-drab colour scheme applied during World War Two. (Author's archive)

Participation in Disaster Relief Operations

On 29 January 1950, for the first time, IIAF used a C-47A for a humanitarian relief operation in support of the people of Zahedan after an earthquake in the east of Iran. On that day, the Dakota airlifted a doctor and representative of the Iranian Red Lion and Sun Society, together with 100 blankets, medicine, bandages and penicillin, from Tehran to Zahedan.

On the following day, another Dakota was assigned to carry three tons of blankets, medicine, tents, shoes and clothing for people affected by the earthquake in Kangan, Bushehr in south-west Iran. It returned to Tehran the same day.

On 6 February 1949, two C-47As, including one carrying Prince Alireza Pahlavi (brother of the Shah) airlifted 3½ tons of clothing, shoes, tents, blankets, medicine and food. One of the Dakotas stayed in Zahedan for three days while the other stayed for four days. A third Dakota flew to Zahedan with tents and carpets onboard on 9 February and returned the next day.

On 29 October 1950, the Iranian Red Lion and Sun Society sent a shipment of humanitarian aid to Pakistan via a C-47. The shipment was handed over to the Pakistani Red Cross to be given to the people in flood-affected areas of Lahore.

Every February and March, the C-47 squadron was tasked to airdrop food and fuel supplies for the villages affected by heavy snowfall due to road blockages. The food supplies helped villagers in mountainous regions of Alborz and Zagros survive the harsh winters and prevented the death of their livestock as well as wildlife.

In addition, the C-47s were used to airdrop aid for train passengers trapped by heavy snowfall in the mountains. For example, on 7 February 1949, a train on its way from Tehran to Miyaneh in north-west Iran was stopped on its journey after the railroad was blocked by snow. Passengers were stranded in the mountains so, to help them survive until rescue teams arrived and/or the snow melted, the IIAF air-dropped several tons of food and fuel.

In use of the Military Air Transport Service

From March 1949 until December 1950, C-47As airlifted 2,809 passengers and 68,816kg of freight in 635 flying hours. In summer 1950, they were used to transport IIAF officers and their families to summer camp in Babolsar on the coast of the Caspian Sea. They were used in several disaster relief operations and were also used to transfer athletes and army officers abroad. Decisions were made to include these flights as a part of a Military Air Transport Service (MATS).

On 29 January 1951, the IIAF established the Military Air Transport Service with six C-47A Skytrains with s/n 5-02, 5-03, 5-04, 5-05, 5-06 and 5-07. A ceremony was held at Ghaleh-Morghi airport, with high-ranking commanders of the Iranian Armed Forces and several members of US Military Assistance Advisory Group (MAAG) in Tehran present. Three of the USAF's non-commissioned officers and three US MAAG advisors received awards from the IIAF for their work in creating the Military Air Transport Service.

Before and during the Iranian New Year holidays (13 March and 8 April 1954), the IIAF's MATS airlifted 343 military officers and their relatives and their belongings. In addition, 751kg of additional cargo was carried from their bases to their home towns by the Dakotas of the MATS.

The IIAF's C-47As were painted white with their undersides bare metal after their overhaul or depot maintenance at 'Siah' (Black) maintenance hangar in Doshan-Tappeh airfield during the 1950s. (Author's archive)

In 1955, IIAF's MATS had following monthly flights:

Route Number 1: Tehran–Tabriz–Kermanshah–Khorram Abad–Tehran
Route Number 2: Tehran–Ahvaz–Abadan–Shiraz–Isfahan–Tehran
Route Number 3: Tehran–Isfahan–Kerman–Zahedan–Khash–Mashhad–Tehran
Route Number 4: Tehran–Kerman–Bandar Abbas–Jask–Chabahar–Khash–Zahedan–Kerman–Tehran
Route Number 5: Tehran–Shiraz–Noshahr–Bandar Lengeh–Shiraz–Tehran

Deadly Crash of the Dakotas

In 1951 and 1952, the IIAF lost two of its C-47As during night-training flights resulting in the death of crew members of both aircraft.

The first accident happened to s/n 5-04 on 21 November 1951. The aircraft, under control of pilot Captain Jaafar Ghefrani, co-pilot Captain Zareh Loukasian, radioman Chief Warrant Officer Qodratollah Talischi and Technician First Sergeant Ismail Pakzad crashed at 19:00 local time when it was departing Mehrabad International Airport for instrument flight training at night. Zareh Loukasian was an Inspector of the Mixed Aviation Wing at that time. Student pilot Sergeant Major Rezakhan Modarres, was the fifth person to lose his life in the crash.

Almost two months later, the IIAF lost another Dakota and its crew on 22 January 1952. The C-47A with s/n 5-02 crashed during an instrument flight training at midnight in south-west Alishah Avaz district near Tehran. It had been flown from Mehrabad International Airport and was under control of Major Khalil Edalat (pilot) and 1st Lieutenant Alimohammad Javanshir (co-pilot). The other people onboard were 1st Sergeant Mohammad Mansouri (engineer), 3rd Sergeant Ismail Bakhtiari (engineer), 3rd Sergeant Hasan Khosh-Tinat (radioman), and 3rd Sergeant Houshang Mostafa Qoliyan (radioman). They all lost their lives.

The IIAF lost at least three more C-47As in other accidents. On 20 June 1961, an aircraft collided with a mountain while dropping skydivers during an exercise. Nine air force and army officers lost their lives.

The next known accidents took place on 11 December 1962 and 30 September 1964. Details are unknown, only that the last accident led to the death of one crew member near Shiraz on the same day.

After the deadly crashes of 1951 and 1952, the IIAF sent its Dakota pilots to Dhahran Air Base to train for instrument landing in bad weather conditions, day and night. On 3 May 1953, the IIAF sent two Dakotas with s/n 5-03 and 5-07 to Dhahran in a flight lasting four hours and 20 minutes. Eight Dakota pilots onboard the aircraft practised instrument landing by means of the ground-controlled approach (GCA) system of the US Air Force Base. Each training sortie in Dhahran lasted three to five hours, during which several pilots were trained on the GCA navigational aid system. After three days, they returned to Mehrabad International Airport.

Expansion of the Dakota Fleet

The loss of the two C-47As in 1951 and 1952 impacted the activities of the Dakota squadron of the Mixed Aviation Wing. It took four years until the US supplied the IIAF with seven more C-47s. They were delivered from October 1956 until October 1957 and received s/n 5-08 to 5-14. With the help of the newly delivered C-47As, the IIAF could spray pastures to combat plant pests. For this purpose, the 'Ashiyaneh Siah' (Black Hangar) responsible for depot-level maintenance of the Dakotas in Doshan-Tappeh airfield designed a special spraying system and installed it under some of the Dakotas.

Despite the delivery of new aircraft, the Dakota squadrons reduced their activities in 1957. In 1955 and 1956, the C-47As flew for 2,795 and 2,204 hours respectively while they flew for only 773 hours in 1957. The cause of the decline was an accident that grounded the fleet for a number of months.

In 1962, the IIAF was supplied with 12 ex-USAF C-47As which were delivered over 18 months. 5-23 was one of them, which was operated from Tehran's Mehrabad International Airport where the IIAF based one of its two Dakota squadrons. (Author's archive)

Several DC-3s, DC-4s and other civilian and military aircraft, including at-least five C-47As are parked at Mehrabad Airport in 1955. (Pakzad, the Iranian National Library)

In 1962, at the request of the Iranian government, the US supplied the IIAF with 12 ex-USAF C-47As, which were delivered between November 1962 and March 1963. The first arrived on 6 November 1962. Brigadier General Zarrabi, commander of the Material Department of IIAF, and Colonel Jack Hughes, the head of US ARMISH in Iran, were present during the handover ceremony.

The new aircraft received s/n 5-16 to 5-28. One of them had a short life and was lost a few days after delivery in an accident on 11 December 1962. As an attrition replacement, another C-47A was delivered in 1963, receiving s/n 5-29. In 1965, the IIAF received its last C-47A plus an EC-47D. They received s/n 5-30 and 5-31 respectively. All of the C-47s were retired from service by 1971, when the IIAF's Fokker F27-400M/600 light transport aircraft had reached complete operational readiness.

Image taken in Shiraz in 1966 shows s/n 5-12, a C-47A delivered in 1957, in use by the Transport Squadron. (Author's archive)

This C-47A with s/n 5-24 is one of two surviving Dakotas of the IIAF today. This example, which is missing a rudder, is kept in the air force museum at Dowshan-Tappeh. (Babak Taghvaee)

Lockheed T-33A Shooting Star: Initiator of the Jet Age

On 29 April 1956, the IIAF entered a new era. The air force received its first jet aircraft, in the form of a pair of former USAF Lockheed T-33A Shooting Star trainers. Almost a year later, the air force began receiving its first jet-engined fighter-bombers, Republic F-84G Thunderjets. Their deliveries were the beginning of a new era in IIAF history and the beginning of the end of piston-engined aircraft. Two years before delivery of the jet-powered aircraft, the IIAF had started a process of retirement and scrapping of its piston-engined aircraft, mainly those that had been left from pre-World War Two days. This decision had been made following the increase in deadly crashes.

In 1960, 1963 and 1965, the IIAF began receiving F-86F Sabre fighter jets, C-130B Hercules tactical transport aircraft and F-5A/B Freedom Fighter fighter jets. In 1968 and 1971, the air force began receiving McDonnell Douglas F-4D Phantom II and F-4E Phantom II multirole fighter jets and RF-4E photo-reconnaissance variants. Beginning in 1957, the IIAF became the most powerful air force in the Middle East and in just 20 years, it grew into one of the most powerful air forces of the world while operating the most advanced combat aircraft and helicopters of that time, such as F-14A Tomcat fighter interceptors and CH-47C Chinook heavy transport helicopters.

The Islamic Revolution of 1979 brought an end to this Persian glory, replacing a secular government with an Islamic regime. The Imperial Iranian Air Force was renamed as the Islamic Republic of Iran Air Force (IRIAF) and faced a significant decline in its power caused by mass early retirements and the imprisonment and execution of its personnel in 1979 and 1980. The Air Force also lost many of its aircraft and hundreds of its personnel, including pilots, in the Iran-Iraq war of the following decade. Financial issues together with negligence as well as sanctions imposed by the United States have so far prevented the government of Iran from rebuilding the Iranian Air Force.

The first jet aircraft of the IIAF, a T-33A previously used by the USAF with s/n 54-1599/TR-599 (c/n 580-9335) at its handover ceremony at Mehrabad International airport on 29 April 1956. This training aircraft received s/n 2-01 and was used by the IIAF for training of F-84G Thunderjet fighter-bomber pilots. (Author's archive)

پیشرفت ایران

شماره ۳۶۰ . این نشریه هر دو هفته یکبار انتشار مییابد

اعلیحضرت همایون شاهنشاه مانورهای اخیر ارتش ایران را که در آن گلوله های حقیقی بکار میرفت مشاهده فرمودند . در این عکس ژنرال جان ساتیز رئیس هیئت مستشاران نظامی امریکا در ایران (سمت راست) و ژنرال گیل هیئت مستشار توپخانه گروه تعلیماتی شماره ۱ در حضور شاهنشاه مشاهده میشوند

نیروی هوائی ارتش شاهنشاهی خود را برای هر جهت آماده میسازد . شش هواپیمای جت تعلیماتی طبق برنامه مساعدت نظامی امریکا بارتش شاهنشاهی تحویل گردیده و است اکنون خلبانان ایرانی مشغول تمرین با آن میباشند

تعلیمات عملی قسمت مهمی از برنامه تعلیماتی افسران و سربازان ایرانی میباشد در این عکس عده ای از دانشجویان آموزشگاه زرهی که در امور مربوط به تانک مطالعه مینمایند در عملیات صحرائی شرکت کرده اند . در دنبال این تانک گروه پیاده و عده ای که با بازوکا مجهز گردیده دیده میشوند . کلیه سلاحها باستثنای تفنگ ها از سلاح های امریکائی میباشند . طریقه تعلیمات نیز مطابق روش جدید و عملیات تاکتیکی هم با شیوه تازه ای صورت میگیرد .

در برنامه مدرن ساختن ارتش شاهنشاهی ایران مستشاران نظامی امریکا فنون جدید ارتش را بافراد ایرانی میاموزند . در مانور سربازان طبق شیوه جدید افراد از زیر آتش گلوله های حقیقی عبور میکنند . این عملیات جزء برنامه تعلیماتی جدیدی است که تحت نظر تیم شماره ۵ انجام میگیرد

از ساعت ۷ تا ۱۰ بعد از ظهر

هر شب برنامه های جالب صدای امریکا بر روی موج کوتاه ۱۳ و ۱۶ و ۱۹ متر و موج متوسط ۲۳۸ متر پخش میشود

ارتش های مدرن امروزه دارای سلاحهای موتوری میباشند و این سلاحها نیروی بزرگ آنها را تشکیل میدهد . امریکا برای تقویت ارتش ایران سلاحهای اتوماتیک و تانکها و سایر سلاحهای موتوری را در اختیار ارتش ایران نهاده است . در این عکس عده ای از دانشجویان دانشگاه نظامی و سربازان تیر با توپ های ۵۵ میلیمتری تیراندازی میکنند .

طبق برنامه مساعدت نظامی و اقتصادی امریکا با ایران تاکنون بیش از ... میلیون دلار مهمات و تجهیزات و لوازم نظامی که وزن آن بیش از ۲۰۰۰۰ تن میباشد بایران تحویل گردیده است . در این عکس اسباب و ابزار یدکی که برای تعمیر و مرمت و نقلیه موتوری ارتش که از امریکا واصل شده مشاهده میگردد

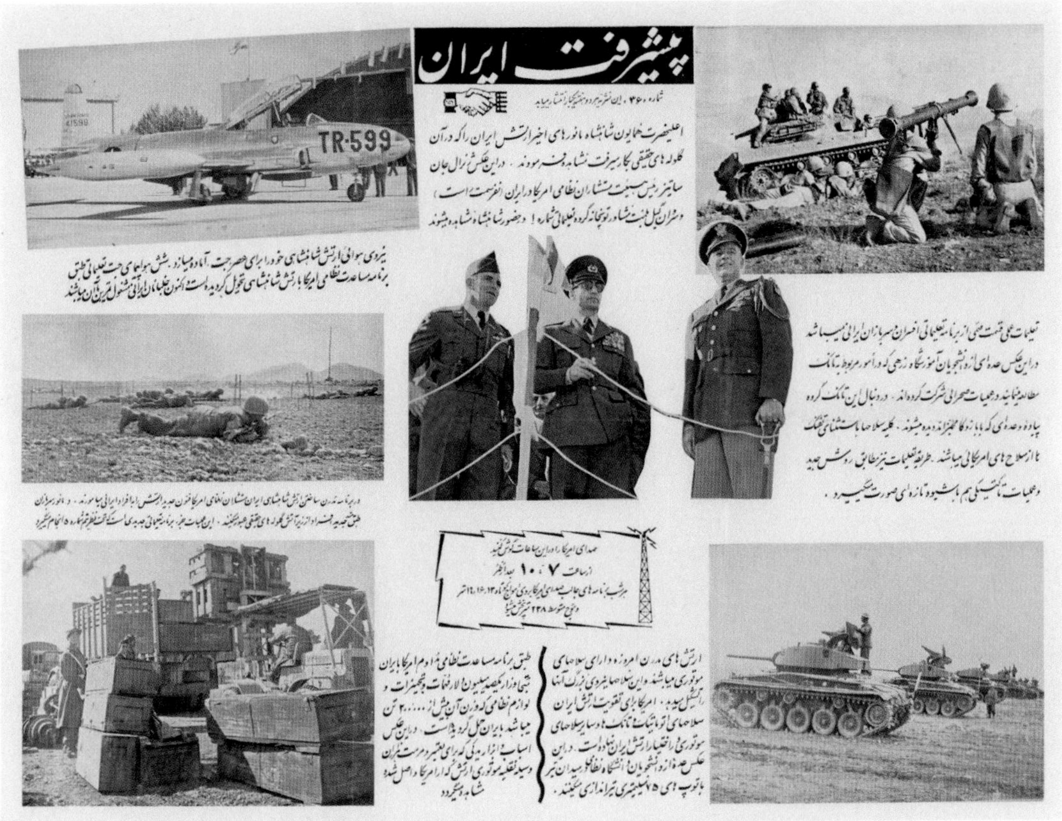

Delivery of the first six T-33A Shooting Stars to the IIAF was a new beginning for Iran's aviation industry. The news about its delivery was published by almost all Iranian newspapers and magazines in April 1956. This example is from Iran *Progress* biweekly bulletin. (Author's archive)

On 15 May 1957, the IIAF received its first F-84G Thunderjet fighter-bombers, which were put in to service with the 1st Fighter Regiment at Mehrabad airport. Just a few days after their delivery, the first of them, with s/n 3-01, flew with two F-47Ds (from the 2nd Fighter Regiment in Ghaleh-Morghi) and a T-33A to Qazvin to test a new radio system that had its antennae installed around Tehran. This image shows two F-47Ds (s/n 2-54 and 2-84) flying with the F-84G. (Author's archive)

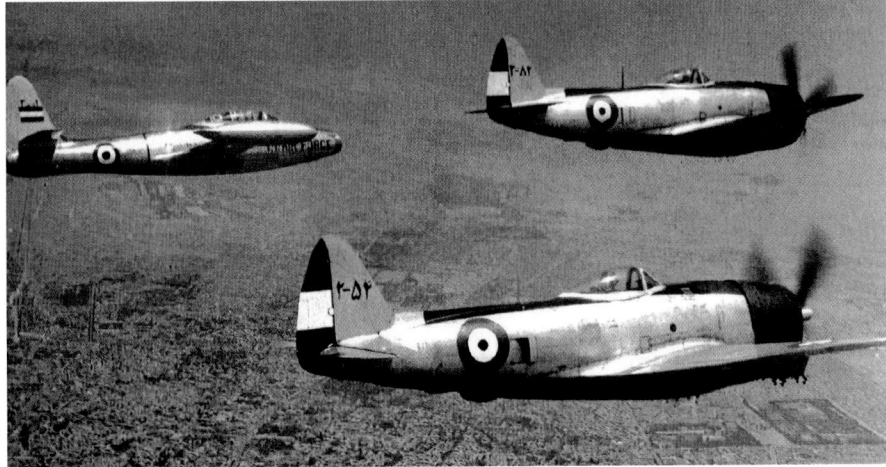

Other books you might like:

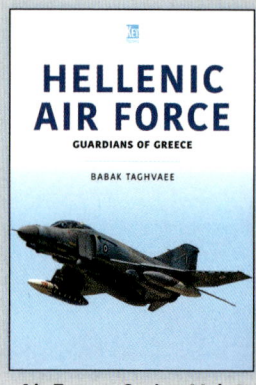
Air Forces Series, Vol. 8

Air Forces Series, Vol. 11

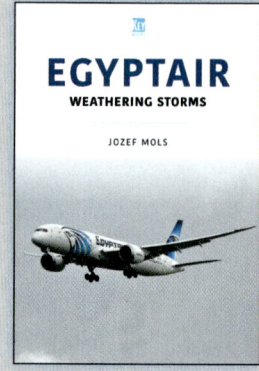
Airlines Series, Vol. 14

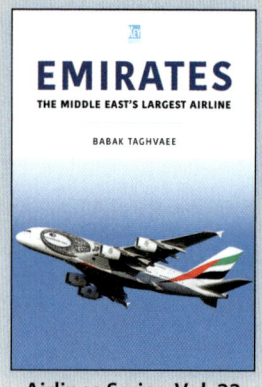
Airlines Series, Vol. 22

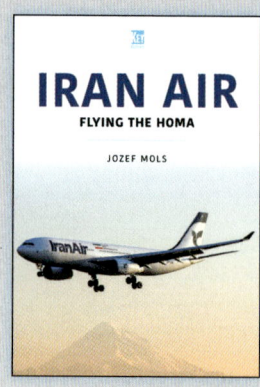
Airlines Series, Vol. 9

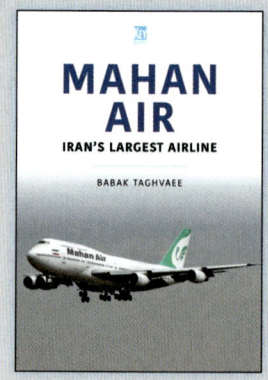
Airlines Series, Vol. 13

For our full range of titles please visit:
shop.keypublishing.com/books

VIP Book Club

Sign up today and receive
TWO FREE E-BOOKS

Be the first to find out about our forthcoming book releases and receive exclusive offers.

Register now at **keypublishing.com/vip-book-club**

Our VIP Book Club is a 100% spam-free zone, and we will never share your email with anyone else. You can read our full privacy policy at: privacy.keypublishing.com